GREAT UNSOLVED MYSTERIES OF SCIENCE

JERRY LUCAS

BETTERWAY BOOKS
Cincinnati, Ohio

Also by Jerry Lucas:
Becoming a Mental Math Wizard

Text illustrations by Valerie Felts
Typography by Park Lane Associates

97 96 95 94 93   5 4 3 2 1

Library of Congress Cataloging-in-Publication Data

Lucas, Jerry, 1950-
      Great unsolved mysteries of science : from the end of the dinosaurs
to interstellar travel and life on other planets / Jerry Lucas. -- 1st ed.
          p. cm.
      Includes index.
      ISBN 1-55870-291-1 : $9.95
      1. Science--Miscellanea.   I. Title.
Q173.L918   1993
500--dc20   92-39006
                                                                              CIP

Acknowledgments

First, I would like to thank the entire staff of Betterway Publications. I have truly enjoyed working with all of them to make this book a reality. Particular thanks go to Susan Morris, whom I have enjoyed working with as my editor on this book and on my previous book, *Becoming a Mental Math Wizard*, also published by Shoe Tree Press.

I would like to thank James Wang, Jim Stein, Scott Wedel, Paul Seah, John Reilly, and Ben Chen for the many thought-provoking technical discussions they have had with me.

Robert Klute, Bruce Bell, Gary Walker, and Jim Stein contributed their own time to act as informal book reviewers.

Sandra Nilluka, Alaska native, provided valuable technical consultation for the material in Chapter 5, as did Fran Seager-Boss, of the Cultural Resources Division, Matanuska-Susitna Borough, Palmer, Alaska.

Special thanks to Sheryl Kelly for her outstanding work in preparing the manuscript for publication, and to Valerie Felts for her likewise outstanding work as technical illustrator.

Special thanks to Fred Reisert of Canisius High School, Buffalo, New York, my former science teacher, for providing me with the motivation necessary to pursue scientific ideas.

Affectionate and loving thanks to my wife Kerry, whose unswerving enthusiasm and encouragement helped to bring this book to completion.

Finally, a very special thanks to my parents, Mr. and Mrs. Gerard R. Lucas, currently of Williamsville, New York, who did so much to encourage my development in science.

Contents

The Facts of the Case
The Impact Theory
Volcanic Eruption Theory
The Continental Drift Theory
Conclusions
Bibliography

The Basics of Weather
Glacier Formation and Movement
Piecing Together the Earth's Climate History
Ice Age Theories
Conclusions
Bibliography

Basic Physics of Motion
Concepts in the Special Theory of Relativity
Propulsion Systems
Other Obstacles to Overcome
Conclusions
Bibliography

The Drake Equation—The Big Picture
Rate of Star Formation (R_*)
The Fraction of Stars that Have Planets
Number of Planets Possessing a Livable Environment
The Likelihood of Life Developing
The Chance of Intelligent Life Developing
The Lifetime of an Intelligent Species
Wait a Minute! What about UFO's?
Putting It All Together
Bibliography

Introduction

Everybody loves a good mystery, but certainly not everyone loves science. It is this book's purpose to show that this unfortunate state of events is a contradiction, since solving mysteries lies at the very essence of science.

Is the Earth really flat? How far away is the sun? Why do objects fall to the Earth? Why is the sun yellow? Can spiders hear? How old is the Earth?

Simple mysteries, posed as these kinds of questions, have led to scientific breakthroughs. People such as Albert Einstein, Sir Isaac Newton, and Galileo brought about such breakthroughs by relentlessly pursuing the answers to intriguing mysteries.

Today, modern science is faced with hundreds of unsolved mysteries. The challenge in writing this book was not to come up with a list of six, but to choose which six out of the many should be included. The solution to any one of these mysteries would constitute a major scientific breakthrough, comparable to those accomplished by Einstein, Newton, and Galileo.

In writing this book, I have striven to make it understandable to anyone with a background in seventh grade general science. At the same time, I believe readers of any background will be sufficiently challenged and intrigued by the material presented here. For readers who wish to explore the subjects in more depth than I have presented, each chapter is followed with a bibliography of more detailed works on the subject.

No greater reward could come to me for writing this book than for one of its readers to someday make a key contribution toward solving one of these great mysteries.

Mysteries and How to Solve Them

For purposes of this book, we will define a mystery quite simply as any question to which we don't know the answer. As such, mysteries abound in our everyday lives.

Who will win the next Super Bowl? What does Suzy Jones think about Charlie Barnett? Will I live to be 100 years old? What will so-and-so say when I tell him the news? What kind of raise will I get? How will I do on my next exam? Who really shot President John Kennedy?

These are just a few examples. Mysteries are everywhere. It is probably fair to say that not a day goes by without each of us thinking a little about some mystery.

This book is about mysteries from science. This chapter is about solving mysteries in general. The techniques presented are applicable to all mysteries. Here is a step-by-step procedure to follow in attempting to solve any mystery.

DEFINE THE MYSTERY

A mystery is a question. Before you start an investigation, you must be clear on what you are investigating. You must formulate a question, then direct your efforts toward answering this question.

This step sounds easy, and it is. But it should not be omitted. If you don't take this step, you can become easily sidetracked later. Identify the mystery and phrase the question. It will help to keep your efforts focused.

IDENTIFY YOUR RESOURCES

What is already out there that can help you get a handle on this thing? What information and facts can you use? If you think about it hard enough, you may find a surprising amount of available information.

Suppose you see an airplane overhead. You formulate a question. How high is the airplane and how fast is it going? What information can you use to figure this out?

You can see how big the airplane appears to you. You can hold a ruler at arm's length and see how big the airplane appears on the ruler, and how fast it is moving along the ruler. You can use geometry to estimate the airplane's altitude and its speed.

You also can hear the sound of the airplane if it is low enough. You can hear the sound change pitch as the airplane passes overhead. But if the airplane is traveling significantly fast compared to the speed of sound, the change in pitch will be delayed by an amount that is related to the airplane's speed and its height above you. This is more information.

You might be able to see the airplane's exhaust. It's tough to make use of this information, but, in theory, if you knew the local temperature, air pressure, and humidity, and could guess the type of plane you were looking at, you could use physics to get information about the plane's altitude and its speed based on the pattern of its exhaust and how fast the exhaust dissipated. You also can observe the plane's direction and, maybe, even what kind of plane it is. You might be able to make an intelligent guess as to the plane's origin and destination.

The point of this little discussion is that a lot of information is usually available to you when you are trying to solve a mystery. Maybe you can't use it all, but maybe you can use more than you think. In the airplane mystery, for example, it is unclear how much use the airplane's direction of travel would be in estimating its altitude and speed. It is also doubtful that the information about the airplane's exhaust pattern would be useful in practice.

GATHER THE DATA

Once you have identified your possible sources of information, you need to go out and gather that information. If your mystery is, "Who will win the next Super Bowl?", then educate yourself so much about this subject that you will be able to speak about it with knowledgeable confidence, not false bravado.

If you make a statement like, "There is a 30% chance that the Washington Redskins will win the next Super Bowl," you should understand all the arguments on both sides of this issue. Who are the main contenders? What if some key players get hurt? How can other teams beat the Washington Redskins?

Collect your data in an unbiased, unslanted way. You are a scientist, not a lawyer. You are trying to get to the truth, not prove a point in a court of law. Familiarize yourself with all aspects of the mystery that you are looking at. Make sure you understand all the data that is available, even though it may not agree with your preconceived notions.

ANALYZE THE DATA

Once you have collected your data, you should organize it and analyze it. This step of the process normally requires using some science or mathematics. How you analyze the data depends on what form your data takes. It is difficult to make any general statements about this step of the process.

FORM THEORIES

A theory is a proposed solution to the mystery. It is okay to have more than one theory at this point. If you have more than one theory, it just shows that you are being objective in your search for a solution. You have not narrowed down the scope of your search to give preference to one line of reasoning. Maintaining your objectivity is a critical part of becoming a good scientist.

The first rule about theory formation is this: A theory should not be formulated until all the data has been collected and analyzed. This rule is so fundamental that it is impossible to understate its importance. Forming a theory before all the data has been analyzed is a typical mistake made by amateur sleuths who are over-anxious in their attempts to solve mysteries. If you form a theory before you have analyzed all your data, then what you really have is not a theory but a hunch. You have a preconceived notion or prejudice that is influencing your thinking. Objectivity demands that you analyze all the available data before you form theories.

It is possible that you can form a hunch and be right. Does that make you a genius, or a person with some type of sixth sense? No. It means you were lucky. Regardless of whether your hunch turns out to be right or wrong, you can be justifiably criticized for proposing a solution before analyzing all the data.

A well-known example illustrating this idea comes from the mystery, "Who shot President Kennedy?" The proposed solution, "Oswald acted alone," was formed long before all the data had been collected and analyzed. Thus, this is not a theory but a hunch. It is not my intent here to argue the case for or against this hunch. But those who formed it, whether right or wrong, can be justifiably criticized for forming their hunch before analyzing all the data.

TEST YOUR THEORIES

Sometimes, but not very often, it is possible to design an experiment to prove your theory right or wrong. A theory like, "Whenever I eat scrambled eggs, I get sick" is very easy to test by experiment. A theory like, "My car will last longer if I use super-unleaded gasoline" is possible to test, but you might have to wait several years before you know the answer.

Most theories are untestable by experimental means. Some examples are "Oswald acted alone in shooting President Kennedy" and "Scorpions will become extinct before humans."

For theories that cannot be tested experimentally, we should make sure that they meet certain dictates of logic. The test, then, does not consist of a physical experiment, but is a logical analysis. Specifically, for any theory to be acceptable, it must satisfy three requirements:

(1) It must be consistent with all the known facts;
(2) It must serve to explain the known facts;
(3) It must be complete and leave no questions unanswered.

Consider the little mystery, "Why is the night sky dark?" We are tempted at once to form the theory, "The sun is the source of light for the Earth and the sun is not visible at night." This theory is consistent with the facts, and it explains the facts. Nonetheless, it cannot be considered complete, since there are many unanswered questions. Stars also supply light, and there are trillions of them out there. Why doesn't their combined light illuminate the night sky? If there are stars everywhere, then, wherever I look, I should see one. Then the night sky should be white, because in all directions my line of sight should be obstructed by a star. Why do I see so many holes of darkness?

The pretty little mystery of the night sky, called *Olbers' Paradox*, is tougher than it sounds. Scientists generally agree that the simple explanation, "The sun is not out," is insufficient. A more viable theory is that stars exist in all directions, but, because the speed of light is only 186,000 miles per second, the light from all of them has not yet reached us. Stars live and die in outer space, and not all of them have been alive long enough for their light to reach us.

In the mystery, "Will it rain tomorrow?", perhaps someone forms the theory, "Yes, because my cat is sick today, and whenever my cat gets sick, it rains the next day."

This theory, if we believe the data, is consistent with the facts. But it neither explains the facts, nor is it complete. To explain the facts, a theory must have a logical link to the set of observed phenomena.

This theory can be amended to say that "a lowering of air pressure affects my cat's inner ear and gives him vertigo, making him throw up." If this is part of the theory, then it explains the facts.

But the theory is still not complete. A lowering of air pressure does not always lead to rain. And how do you know that your cat gets vertigo or that his inner ear is affected? Can't other things besides a lowering of air pressure make your cat sick? There remain many unanswered questions. Until the theory can be changed again to answer these questions, it must be considered incomplete and unacceptable.

When a theory becomes a piece of patchwork with a long list of conditions and amendments, it is probably time to reexamine it. It may also be a sign that the theory was formed before all the available facts were analyzed. In any case, if you find yourself always having to amend your theory in some way, you are best advised to do more thinking. It is likely that you are on the wrong track.

FORM CONCLUSIONS

After your theories have been tested, several things are possible. If none of your theories is acceptable, then you must rethink things and form a new theory. If one and only one of your theories is acceptable, then you are fortunate, because you have a possible solution that is unchallenged. At this point, you should recheck your steps to make sure there is not another possible theory you have overlooked.

The most interesting case is one in which you have multiple theories that are acceptable. In other words, they all meet the three conditions of the previous section. Then how do you select one of them?

Science demands that, when more than one theory is acceptable, you are obliged to select the one that is the simplest. That is, you are obliged to select the theory that has the highest likelihood of being true. This principle, called *Occam's Razor,* is a necessary and important ingredient in the mystery-solving process.

Occam's Razor tells us not to get too complicated and fancy with our theories. Find the simplest, most straightforward explanation, and you have the best chance of solving the mystery.

Suppose you wake up one morning and find that your television is turned on. You are the first one in your family to get out of bed and notice this. How do you explain it? Your first theory might be that someone in your family stayed up late watching television, and when she went to bed, forgot to turn it off. Your second theory might be that an external force, capable of telekinetic powers, decided to play a trick on you.

Which theory do you believe? Here, both theories satisfy the three criteria, so both are acceptable. But, by the dictates of Occam's Razor, you are required as a scientist to accept only the first theory as being the more probable and the simpler of the two.

ANNOUNCE YOUR RESULTS

Why not? If you have followed all the rules carefully and maintained your objectivity throughout the process, you have a well thought-out solution to the mystery. You have a very good chance of being right.

Can Earthquakes Be Predicted?

Some people compare a moderate earthquake to airplane turbulence. The vibrations are about the same size. In a moderate earthquake, walls of buildings creak a bit, dishes rattle, and chandeliers swing from ceilings. Blinds bang into windows. When it is all over, chandeliers and window blinds remain swinging for a minute or so. The same sorts of things would happen in an airplane experiencing turbulence, if the airplane had these objects in it.

But other than that single similarity, an earthquake is much different from airplane turbulence. The strongest vibrations in an earthquake are from side to side. There is some up-and-down motion, but it is almost never the dominant motion. Earthquake vibrations are cyclic. They shake you back and forth in repetitive patterns. Earthquakes rarely last longer than thirty seconds, although enormous earthquakes may go on for a couple minutes.

Almost all earthquakes consist of two main jolts. The jolts correspond to two groups of waves, called P (for Primary) waves and S (for Secondary) waves. P-waves travel faster than S-waves, so they arrive first. But S-waves are more powerful. The most frightening moment of any earthquake is the instant that the S-wave hits. The earthquake seems to take a sudden leap in intensity.

P-waves travel in a direction outward from the source of the earthquake. If you are in a building, you can sometimes figure the direction to an earthquake's epicenter by judging the way that the building vibrates in response to the P-waves. In practice, this is difficult because the P-wave vibrations are not always strong enough to shake a building noticeably.

S-waves shake a building back and forth at right angles to the direction of the epicenter. Thus, if an earthquake's epicenter is located to your north, the building you are in will shake in an east-west direction when the S-waves hit. In any moderate or large quake, you will be able to feel this motion and determine its direction.

Your distance to the epicenter can be estimated by using the fact that P-waves travel faster than S-waves. If you are quick and observant, and can estimate the time interval between the first P-wave and the first S-wave, you can estimate your distance to the epicenter.

This is the same principle that can be used to estimate the distance of a lightning strike. A lightning bolt emits light waves (lightning) and sound waves (thunder). Light waves travel at 186,000 miles per second, but sound waves travel only about 700 miles per hour. The distance to the lightning strike can be calculated using the fact that the waves travel at different speeds. Only the time difference between the arrival of the waves needs to be known. See Figure 2-1.

In this diagram, let v_1 denote the velocity of the faster wave and v_2 that of the slower wave. Let d denote the distance from the event to the observer. Let t denote the time difference in arrivals of the waves. Let T be the time required for the faster wave to reach the observer. Then v_1T is the distance from event to observer. We can say:

$$v_1T = v_2(T + t)$$

This just says that both waves travel the same distance. The faster wave travels the distance in time T. The slower wave requires a longer time T + t. We can use algebra to solve for T:

$$T = (v_2t)/(v_1 - v_2)$$

The distance from event to observer is v_1T or:

$$d = v_1T = t(v_1v_2)/(v_1 - v_2)$$

In the case of lightning, v_1 is 186,000 miles per second and v_2 is 700 miles per hour, or about 1/5 mile per second. Since v_1 is so much faster than v_2, we can derive the rule that the time difference in seconds divided by five approximates the distance from the observer to the lightning.

In the case of an earthquake, the velocity of P-waves through rock is roughly 3.5 miles per second, and the velocity of S-waves is roughly 1.9 miles per second. By plugging these values in for v_1 and

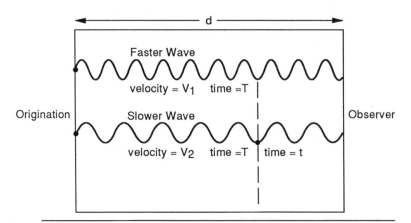

FIGURE 2-1. *Finding the distance to a wave source based on observing two waves. The observer needs to know the speeds of the two waves and the time difference between their arrivals.*

v_2 in the above equation, we can derive an approximate rule that the distance from an observer to the quake's epicenter in miles is about four times the difference in arrival times in seconds between the P-wave and the S-wave. The number four is only an approximation and depends on the types of rocks involved. In reality, this number can vary between two and ten.

In principle, you can make a very good guess at the epicenter of an earthquake by estimating its distance from you (from the time between P- and S-waves) and its direction from you (from the direction that your building shakes in response to the S-waves). Unfortunately, this is not quite enough information. For example, if your building shakes in an east-west direction during the S-waves, you know that the epicenter is either to your north or to your south, but you don't know which.

If your measurements are accurate, you can narrow the location down to one of two points. To get the exact location, you need a second observer at another spot to measure either the distance or the direction from his spot. Then the epicenter can be pinpointed.

Figure 2-2 illustrates a case of locating the epicenter of the 7.1 Loma Prieta earthquake, which struck the San Francisco area on the afternoon of October 17, 1989. An observer in San Jose estimates the distance to the quake's epicenter at forty miles, and the direction as either north or south. Based on the first observer's data, the quake is centered either in the Oakland Hills area (forty miles to the north) or in the Santa Cruz area (forty miles to the south). A second observer, twenty-five miles to the northwest in the town of San Mateo, is able to estimate the direction to the epicenter as on a line from the north-northwest to the south-southeast. The quake could not have been north-northwest of San Mateo. This would have been inconsistent with the data collected by the first observer. But south-southeast of San Mateo, the sightings intersect in the Santa Cruz area. The Santa Cruz area, therefore, is a reasonable guess as to the location of the quake's epicenter.

EARTHQUAKE BASICS

You may have heard of a device called the Foucault pendulum. There is a good chance that your local Museum of Science has one on display. The Foucault pendulum is nothing more than a large mass suspended from a long wire and free to swing in any direction. As the Foucault pendulum swings back and forth, the Earth rotates beneath it. Thus, a relative motion between the Earth and the Foucault pendulum can be observed. In fact, the Foucault pendulum is proof positive that the Earth is rotating.

A modern seismograph, designed to measure the vibrations of the ground during an earthquake, operates on the same principle. A heavy

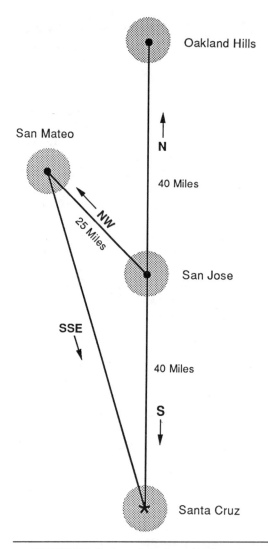

FIGURE 2-2. *Locating the epicenter of an earthquake with two observers. In the diagram, the observer in San Jose estimates the distance away and the direction (either north or south in this case). The second observer, in San Mateo, estimates another direction. There is only one location that is consistent with the data taken by the two observers.*

weight hangs from a support structure that is anchored to the Earth. It is let to stand motionless. When the Earth shakes, the heavy weight on the end of the pendulum will tend to remain motionless while the support structure, anchored to the shaking Earth, moves. Thus, ideally, a relative motion between the Earth and the heavy weight can be observed. (See Figure 2-3.)

In practice, the heavy weight will move some, and a device called a *damper* is used to minimize this motion. Further modernization of this pendulum-type seismograph allows for an electric current to be produced as the heavy weight and the Earth move relative to each other. The current increases in proportion to the size of the relative movement.

As the Earth moves back and forth, a pattern of waves is produced by the seismograph. Each wave has a peak, or high point, and a trough, or low point. The amplitude of the wave is defined as the wave size, or one-half the difference between the peak value and the trough value. The Richter magnitude of an earthquake is defined as the logarithm to the base 10 of the maximum seismic wave amplitude recorded on a standard seismograph at a distance of 100 kilometers from the earthquake's epicenter.

This definition requires that everyone using the Richter scale to

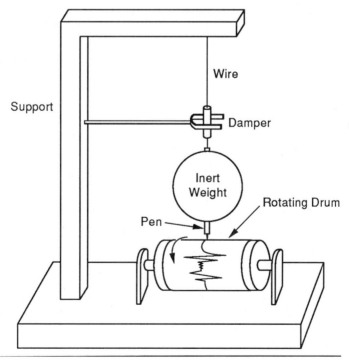

FIGURE 2-3. *Basic elements of a seismograph.*

measure earthquake magnitude must use what is considered a standard seismograph. It does not mean that the standard seismograph must be exactly 100 kilometers from the earthquake's epicenter. Knowledge of the epicenter location and the maximum wave amplitude recorded at any location is sufficient information to compute the Richter magnitude mathematically.

Magnitudes and Geographic Distribution of Earthquakes

Because the Richter magnitude is a logarithmic measure, an increase by one in the Richter magnitude translates into a quake that is ten times more powerful. Thus, a magnitude 8 earthquake is ten times more powerful than a magnitude 7, 100 times more powerful than a magnitude 6, and 1,000 times more powerful than a magnitude 5. This thought is sobering.

The October 17, 1989 quake measured 7.1 on the Richter scale and resulted in the deaths of sixty-seven people. It caused the Bay Bridge, connecting San Francisco with Oakland, to collapse. Many buildings and homes were demolished. Of those people who were not killed or injured, many suffered severe hardships as a result of the quake. Few can imagine a quake ten or more times that size, such as the magnitude 8.3 that struck San Francisco in 1906 or the other magnitude 8.3 that hit Alaska in 1964. The most devastating quake in terms of human cost during the 1980s was the magnitude 8.1 quake that hit Mexico in 1985. Approximately 10,000 people were killed. In 1976, an earthquake in China was estimated to have killed a quarter of a million people.

Earthquakes do not strike equally everywhere. The majority of the world's earthquakes strike in a region called the Pacific Rim or the Ring of Fire, so called because the region also contains a large number of active volcanoes. The Pacific Rim is a narrow band that extends along the west coast of South America, through western Central America and western North America to Alaska. There the band arcs west through the Aleutian Islands and intercepts the Kamchatka Peninsula of Siberia. From there, it heads south through Japan, the Philippines, and Indonesia.

Other areas of high earthquake activity are China, Afghanistan, Iran, Turkey, Greece, Yugoslavia, and Italy. Many earthquakes also occur undersea. A particularly interesting band of active earthquake behavior lies along the so-called Mid-Atlantic Ridge, a range of mountains running along the floor of the Atlantic Ocean from Iceland almost all the way to Antarctica. Earthquake activity is not confined to our planet. Seismic equipment has recorded quakes on both the moon and Mars.

Different Seismic Waves

An earthquake's epicenter is not the true point of origin of the seismic waves. The seismic waves originate at a point called the *focus*, which can be anywhere from on the surface to 400 miles deep. The epicenter is the point on the Earth's surface directly above the focus. Seismic waves spread out in all directions from the focus. They can be recorded by sensitive equipment in other parts of the world after having traveled through the Earth's interior.

P-waves and S-waves are called *body waves* because they arrive from the focus of the earthquake. P-waves are compression waves, and travel in the same manner as sound waves. They alternately stretch and compress the rock, and vibrate in the same direction in which they travel.

S-waves have a motion that is much more complex. They vibrate at right angles to their direction of travel, and also produce some up-and-down movement. They can best be visualized by thinking of a squirming snake crawling over a piece of wood. The snake's body zigzags from side to side, but it must also go up and down to get over the piece of wood.

In addition to the body waves, there are surface waves. Surface waves result from body waves that encounter the surface and are then reflected in such a way as to run along the surface. Surface waves are insignificant for deep earthquakes. In general, the strength of the surface waves decreases as the depth of the focus increases. At considerable distances from the earthquake's epicenter, however, surface waves are usually the strongest of the waves. This is because they benefit from reflection of P-waves and S-waves at many points along their paths. This factor slows their rate of decay appreciably as the distance from the epicenter increases. Love waves are the faster moving surface waves. They whipsaw the ground back and forth in the same manner as S-waves (at right angles to their direction of travel), but they don't have the vertical movement of the S-waves. Rayleigh waves are slow, rolling waves that move up and down as well as back and forth. Unlike S-waves and Love waves, they vibrate in the same direction that they travel.

In almost all cases, earthquake waves are silent. The lowest wave frequency audible to the human ear is about twenty Hertz, or twenty cycles per second. Almost always, earthquake waves vibrate at lower frequencies in what is termed the *infrasonic* range. Therefore, all the sounds you are likely to hear during an earthquake are from the vibrations of objects such as walls of buildings. Claims that animals can sense the coming of an earthquake before it arrives have not been verified scientifically. If these claims are true, the unusual animal behavior is most likely due to gases escaping from the ground just prior to an earthquake, or possibly to electrical phenomena that sometimes accompany earthquakes.

Locating an earthquake's epicenter is located by calculating the difference of arrival times between P-waves and S-waves at different stations. In theory, a station can use the S-P time to compute its distance from the epicenter. The epicenter lies somewhere on a circle with its center at the station and its radius the distance to the epicenter from the station. Similar circles at two other stations will intersect at one point, the epicenter. (See Figure 2-4.)

Modern installations use sophisticated computer equipment configured in an array to chart the arrival of earthquake waves. Each station in the array has a clock synchronized with all other stations. Arrival times of waves at each station can be recorded and compared with those at the other stations. Thus, a huge amount of information about waves and their speeds can be learned.

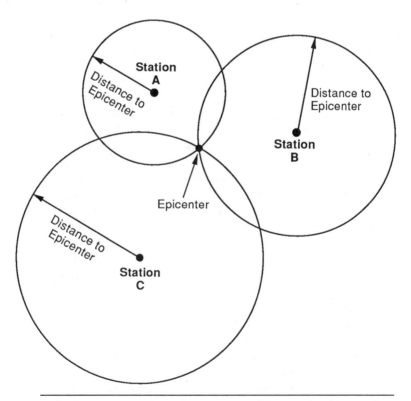

FIGURE 2-4. *An earthquake's epicenter can be located by three observers, each knowing only the distance from his station to the epicenter. Three circles intersect in a single point, which defines the epicenter.*

WHAT CAUSES THE WAVES?

Imagine trying to balance your ballpoint pen so that it stands on its tip. If your pen is perfectly uniform in its weight distribution, it is theoretically possible to do it. But, realistically, you will never be able to do it. Why? When the pen is balanced in this position, it is termed to be in *unstable equilibrium*. This means that any force that acts on the pen can disturb it from its position, and the pen cannot recover. You can never hope to achieve balancing the pen like this because you cannot as a human being balance all the forces so precisely. Even if you could, a slight warp in the table or a tiny air current would be sufficient to topple the pen.

Now imagine the chair on which you are sitting. If you stand up and give it a slight push, it will vibrate back and forth, eventually coming to rest in its initial position. The chair is in stable equilibrium. If you or anything else exerts a force on it, it will eventually return to equilibrium. This assumes, of course, that the force is not so large as to knock the chair over completely. Objects in stable equilibrium have built-in restoring forces to help them return to their equilibrium positions if they are disturbed.

Simple Harmonic Motion

Hooke's Law states that objects disturbed from stable equilibrium will experience a restoring force that is proportional to the amount of displacement and acting in the opposite direction. An example is a mass being suspended from a spring. When the spring is released, a restoring force takes effect. The more the spring is stretched, the stronger will be the restoring force, according to Hooke's Law.

After the mass is released, it will go to its equilibrium position and past it because of its momentum. Then the restoring force will begin operating in the opposite direction, but always toward the equilibrium position. The mass will pass the equilibrium position again, descend to its initial displacement, and the restoring force will again pull it up.

If there were no friction in the system to slow things down, it would vibrate back and forth forever in this manner. This type of vibration is called simple harmonic motion. It is of tremendous practical importance in physics, since, by Hooke's Law, objects disturbed from a stable equilibrium position experience a restoring force that is proportional to their displacements, and thus respond to the disturbance with simple harmonic motion. Simple harmonic motion can be described mathematically by trigonometric functions. The location of a particle undergoing simple harmonic motion, when graphed against time, produces a wave function. (See Figure 2-5.)

Examples of simple harmonic motion include electrons in a wire carrying alternating current, vibrating air molecules carrying a sound

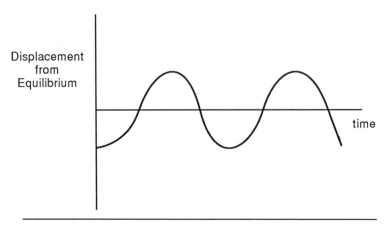

FIGURE 2-5. *An object describing simple harmonic motion defines a wave function as its displacement from equilibrium is graphed against time.*

wave, and motions of vibrating parts of musical instruments. Simple harmonic motion is applicable to earthquakes. A mass of rock displaced from an equilibrium position is very analogous to a mass hanging from a spring. When the force holding the rock in its displaced position is released, the rock will experience a restoring force (according to Hooke's Law). It will vibrate with simple harmonic motion in an attempt to reach equilibrium. Since there is friction in the system, the vibrations will eventually cease.

Creation of Waves

Any vibrating object generates waves. As the mass attached to the spring vibrates up and down, it generates compression waves in the spring. The spring goes through alternate periods of compression and expansion as the mass moves up and down. Similarly, we could imagine a stiff rope attached to the side of the mass and anchored at its opposite end, which is far away. As the mass moves up and down, it will whip the rope, creating transverse waves. Transverse waves differ from compression waves in that they vibrate at right angles to their direction of travel. Compression waves always vibrate in the same direction as they travel.

The example of the mass suspended on a spring is one-dimensional. The motion of the mass is up and down, in only one dimension. The mass is treated as a point, not as a complex geometric shape.

As masses of rock underground vibrate to produce earthquake waves, the situation is far more complex. The mass of rock is not a mere point, but a complex geometric shape in three dimensions. The vibrations it undergoes take place in three dimensions. But the basic principles are the same as those of the simple mass on a spring. Vibrating rocks create both compression waves and transverse waves, or P-waves and S-waves.

Reflection and Refraction

Wave motion is further complicated by two phenomena of behavior, called reflection and refraction. When a wave encounters a barrier, it will reflect off the barrier at the same angle at which it struck the barrier. In words, this law says that the angle of incidence equals the angle of reflection. Anyone who has looked in a mirror has observed this phenomenon as it applies to light waves.

When a wave encounters a boundary, it loses energy to the boundary. It exerts a force on the boundary, which starts the boundary itself vibrating. In the case of light waves bouncing off a mirror, this effect, of course, is negligibly small. But in the case of earthquake waves encountering the surface of the Earth, the effect is significant enough to produce Love waves and Rayleigh waves, as discussed earlier in this chapter.

A non-wave example illustrates the principle. A billiard ball striking the cushion of the billiard table will reflect off the cushion at the same angle at which it arrived. (See Figure 2-6.) But it will lose a little speed as it comes off the cushion because some of its energy will be imparted to the cushion. Although difficult to measure or observe, the particles on the cushion will vibrate with simple harmonic motion, and waves will propagate down the length of the cushion.

A phenomenon known as refraction occurs when waves pass from

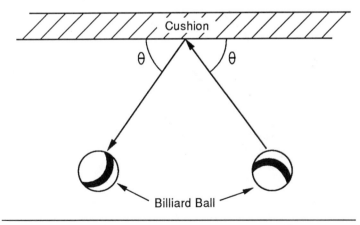

FIGURE 2-6. *A billiard ball striking a cushion bounces back out in such a way that its angle of reflection equals its angle of incidence. This assumes that the ball does not have a rotational spin or "english" imparted to it.*

one material to another. This is because the speed of the waves depends on the properties of the material they are passing through. The effect of refraction causes waves to turn as they pass from one material to another at an angle. For example, in the case of a wave front leaving one material and entering another where its speed is slower, the wave slows down and changes direction. The new wave direction is closer to the perpendicular of the boundary.

This may sound complex, but it is actually a direct result of geometry. A simple analogy is to imagine a row of soldiers walking on a smooth grassy field, who suddenly encounter a large area of mud. If they encounter the muddy area at an angle, the first soldiers to reach the mud will slow down first. When all soldiers have reached the muddy area, their direction will have changed. (See Figure 2-7.) Refraction has application to the study of seismic waves. As waves pass through the Earth's interior, they are refracted as they pass through several layers of the Earth's core. Roughly speaking, the Earth's interior consists of a mantle, a molten outer core (liquid), and an inner core. P-waves are refracted as they encounter each of these barriers, and the amount of refraction can be observed. From these observations, information about the composition of the Earth's interior can be deduced.

A similar interesting deduction can be made since P-waves can pass through water and S-waves cannot. As S-waves were never ob-

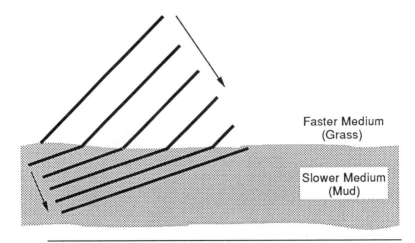

Faster Medium
(Grass)

Slower Medium
(Mud)

FIGURE 2-7. *Rows of marchers moving from a grassy (higher speed) area to a muddy (lower speed) area experience a change of direction at the boundary. This effect is called "refraction," and is a common property of waves. Light waves exhibit the same effect as they move from air to water or from hot air to cold air.*

served to penetrate the Earth's molten outer core, geologists thus concluded that this part of the Earth's interior is liquid.

Related Mathematics

The mathematics of wave propagation are expressed by the wave equation, a complex idea of higher mathematics. The wave equation and how to solve it are beyond the scope of this book, but we will discuss it in general terms.

The main underlying assumption of the wave equation is Hooke's Law. From there, a complex equation is derived that relates positions of particles with respect to time. The equation is referred to as a *partial differential equation*. It is called this because it involves rates of change of position with respect to time. Rates of change in mathematics are often referred to by the term *differentials*.

In one dimension, the wave equation is commonly called the *vibrating string equation*. It is the simplest form of the wave equation. In two dimensions, the wave equation is used to analyze vibrating membranes such as bongo drums. It is considerably more complex than the vibrating string equation. The three-dimensional wave equation is more complicated still, and is used to describe earthquake waves. Solutions of differential equations such as the wave equation are not numbers, but functions. The functions describe motions of particles moving in time as a function of time. A complete solution to the three-dimensional wave equation involves multiple solutions superimposed or added together. These solutions can be grouped into two general classes, one class corresponding to the P-waves and one class corresponding to the S-waves.

THEORY OF PLATE TECTONICS

The theory of plate tectonics is one of the most revolutionary ideas to come out of twentieth-century physics. It suggests that continents are not stationary but move, or, more accurately, float. The surface layer of the Earth, called the lithosphere, slides on an underlying layer of partially molten rock called the asthenosphere. The lithosphere consists of a set of plates that move along atop the asthenosphere, colliding with each other along boundaries. This is the essence of the theory of plate tectonics.

The implications of the theory are far-reaching and remarkable. It helps explain many mysteries such as "How are mountains formed?" or "Why do fossils of marine animals lie on mountaintops?" Drifting continents affect life on Earth and the evolution of life. The colliding of two continents can result in the extinction of one species and the survival of another. The poleward drifting of a continent on which animal species are stranded will favor the survival of only those species that can adapt to the resulting climatic change.

The theory of plate tectonics also has application to earthquakes. For it is along the boundaries of the plates that nearly all earthquakes occur.

Evidence for Plate Tectonics

Many rocks act as strong magnets and are affected by the Earth's magnetic field. As these rocks are buried by sediment, they become historical compasses. They align themselves in the direction of the Earth's magnetic field at the time of their creation.

Studies of rocks from the same parts of the world but of different ages have yielded interesting results. The rocks from different ages were aligned differently, as if the location of the Earth's magnetic north pole was shifting through time. In fact, this was the first, and perhaps the most natural, explanation at the time for these observations. But further studies of rocks on other continents led to inconsistencies. Rocks in England of a certain age would point toward a different north pole than rocks in North America of the same age. How could this be explained? There was only one explanation. The continents themselves had moved. It was an inescapable conclusion, the only one consistent with all the available facts.

Types of Plates

Alfred Wegener, the proponent of the theory of plate tectonics, was originally a meteorologist. He spent a good deal of his time in Greenland, studying climate conditions and climate history there. It is often said that he was inspired to his theory of plate tectonics by gazing at giant icebergs floating in the ocean.

The comparison is interesting. Both icebergs and tectonic plates are affected by the force of buoyancy. The physical law governing buoyancy behavior is called Archimedes' Principle. One way to state Archimedes' Principle is that when an object is immersed in a fluid, it displaces an amount of fluid that is equal to its weight. Thus, an iceberg lies approximately nine-tenths underwater and one-tenth above water. This is because the iceberg is less dense than liquid water. Nine cubic feet of liquid water will weigh about the same as ten cubic feet of iceberg. When a ten-cubic-foot iceberg is immersed in water, by Archimedes' Principle it will displace nine cubic feet of water. Then one cubic foot of iceberg will be left to stick out of the water. Objects that are more dense than water will sink.

Tectonic plates are immersed in the asthenosphere, a fluid consisting mostly of molten rock. How low or high the plates float on the asthenosphere depends on their densities. Plates that are much lighter than the fluid in the asthenosphere will ride higher than plates that are only slightly lighter. Plates that are heavier will sink through the asthenosphere into the Earth's core, and will no longer be plates.

The difference between continental and oceanic plates is their densities. Continental plates are less dense than oceanic plates. They therefore ride higher on the underlying fluid of the asthenosphere. The Earth's water then settles into the lower lying oceanic plates. Technically, the boundaries between the continental plates and the oceanic plates do not lie exactly along coastlines. There is some overlap, with pieces of continents lying on oceanic plates, and vice versa. But, to a close approximation, this simple model explains the distribution of the Earth's oceans.

INTERACTIONS BETWEEN PLATES

In all, there are about a dozen major plates and several smaller ones. At plate boundaries, activity occurs, which leads to earthquakes. This type of activity depends on the nature of the plate boundaries. There are three different classes of plate boundaries: divergent, convergent, and transform.

Divergent Plate Boundaries

A divergent plate boundary occurs where two plates are moving away from each other at their boundary. (See Figure 2-8.) When this happens, a gap is created between the plates. Magma from the Earth's interior wells up into the gap, forming more material for the plates. The plates then grow in area, since the magma adds material to them. This creates stresses in the rock. Mild earthquakes occur to either side of the plate boundary as the existing material in the plates is squeezed.

Most divergent plate boundaries occur on ocean floors. The Mid-Atlantic Ridge, which runs from Iceland to the Antarctic Ocean, is actually a long divergent plate boundary. In the North Atlantic, the ridge forms the boundary between the North American Plate and the Eurasian Plate. Farther south, the ridge forms the boundary between the North American Plate and the African Plate. Still farther south, the plate marks the boundary of the South American Plate and the African Plate.

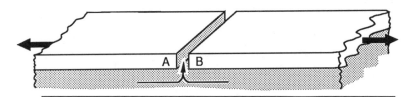

FIGURE 2-8. *A divergent plate boundary. Magma wells up through the cracks that are created as plates A and B move apart.*

Studies have been made on the sedimentary rocks lying on the ocean floor on either side of the Mid-Atlantic Ridge. These rocks were formed as magma from the ridge cooled and hardened. The youngest rocks are the ones closest to the ridge. This observation is an interesting confirmation of seafloor spreading, which is caused by extra plate material being formed at a divergent boundary. An extension of this idea is to theorize that at one time South America and Africa were joined together. The two continents fit together only fairly well, but the fit is very good if taken to a depth of about 3,000 feet, the approximate depth of the continental shelves. In fact, we are now fairly certain that South America and Africa once fit together. This is validated by the reptile Mesosaurus, which lived only in fresh water about 300 million years ago. The fossils of Mesosaurus can be found only in Africa and South America.

Besides in the middle of oceans, divergent plate boundaries also exist over continents. The most famous example is the Great Rift Valley of Africa. It extends approximately 2,500 miles from Ethiopia to Mozambique. This remarkable area contains many trapped lakes and volcanoes. Volcanoes testify to the magma welling up to fill the gap between diverging plates. This magma contains basaltic material, which is the stuff typical of oceanic crust. It is denser than continental crust, and rides lower atop the asthenosphere. Thus, this area is a deep valley, and is actually an ocean in the making. The Rift Valley is slowly tearing Africa apart, just as the Mid-Atlantic Ridge, millions of years ago, tore Africa and South America apart.

Convergent Plate Boundaries

Convergent plate boundaries occur where plates collide head-on. Convergent plate boundaries consist of three general classes: both plates involved are continental plates; one plate is continental and the other is oceanic; or both are oceanic.

If both of the colliding plates are continental, then the plates are of approximately equal density. They smash into each other head-on. Mountains are created by this process, as the plates fold like an accordion when crushed together. Iran, Turkey, the Alps, and the Himalayas are the most dramatic examples of this mountain-building process still in progress. Numerous earthquakes are caused by the resulting stresses on the rocks as the plates crumple together.

If one of the plates is continental and the other is oceanic, the denser oceanic plate will dive below the lighter continental plate in a process called subduction. (See Figure 2-9.) Subduction ensures that the total amount of crust is conserved. New crust is being created by upwelling magma at divergent plate boundaries. The resulting excess crust must be recycled somehow. The answer is in subduction. The denser oceanic plate dives beneath the continental plate, where it

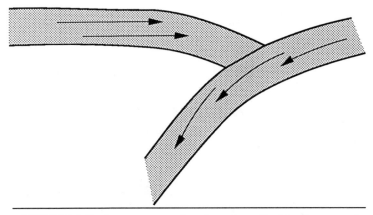

FIGURE 2-9. *A convergent plate boundary. In this diagram, the heavier oceanic plate (right) dives below the lighter continental plate (left) in a process called subduction.*

eventually melts and is recycled back into the mantle. Thus we have a sort of "crust cycle," in which magma from the mantle creates new crust at divergent boundaries, but is eventually subducted and then recycled again into the mantle.

Ocean trenches form in subduction zones. One dramatic example is the long ocean trench that extends along the entire west coast of South America. It is formed by the oceanic Nazca Plate subducting under the continental South American Plate.

Earthquakes occur at boundaries, marked by the ocean trenches, as the plates grind past each other through the subduction zone. More earthquakes occur as the subducting plate descends farther down and meets the asthenosphere in an area known as the Benioff Zone. These earthquakes are likely due to deformation and resulting stress as the cool subducting rock collides with the hot molten material of the asthenosphere.

Thus, along the South American coast, a first group of earthquakes occurs offshore along the ocean trench. This is where the Nazca Plate grinds past the South American Plate during subduction. A second group of earthquakes occurs under the South American continent, as the subducted plate enters the asthenosphere, passing through the Benioff Zone.

If both plates are oceanic, the heavier, denser plate will dive below the lighter plate, creating a subduction zone. The result is both a trench and a neighboring arc of volcanic islands. We will discuss this case more in the next chapter. In this case, earthquakes occur along the trench and along the island arc, as the lower plate dives into the asthenosphere. Examples of this type of plate interaction are the

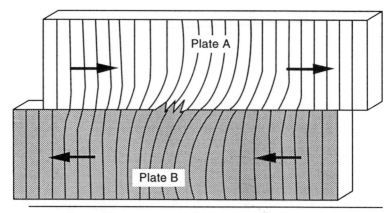

FIGURE 2-10. *A transform plate boundary. Plates A and B slide past each other, creating horizontal displacements in adjoining landforms.*

Aleutian chain of Alaska and the Marianas of the western Pacific.

Transform Plate Boundaries

A transform plate boundary exists where two plates are moving past each other along their edges. (See Figure 2-10.) In reality, this category is a simplification, since no two plates move past each other without at least a small bit of subduction or divergence occurring.

Earthquakes occur along this type of boundary as the rocks move past each other. An example is the San Andreas Fault of California, where the Pacific Plate is sliding past the North American Plate. In some respects, earthquakes along transform faults are the most interesting, since their effects can be most easily observed by people. Horizontal displacements of fences, road surfaces, and other objects are typical results of earthquakes at transform faults.

Putting It All Together

At points where two plates come together, tension is created. Plate A wants to move somewhere but can't because of resistance or friction provided by Plate B. Such is the situation at convergent or transform plate boundaries. At divergent plate boundaries, a plate wants to move in a certain direction but can't because of its own internal stresses.

In all cases, an analogous situation can be seen in hanging a mass on the end of a spring and stretching the spring. The mass wants to move up but can't because of the resistive force you are applying.

Then something gives way. Stresses build to the point where the resistive force can't resist any more. The rocks slide along, or your hand lets go of the mass on the end of the spring. In both cases, the response is simple harmonic motion, producing waves. Since friction exists, the simple harmonic motion gradually comes to a stop. That is what an earthquake is.

THEORIES OF EARTHQUAKE PREDICTION

No earthquake would happen without water. Laboratory experiments have proven this. When rocks containing no porous water are subjected to stress in the laboratory, they do not react by fracturing in a manner consistent with earthquake occurrence.

This curious fact is validated in part by the increased number of earthquakes occurring near manmade reservoirs and dams in seismically active regions. Examples include the Hoover Dam of Arizona and the Lake Kariba reservoir in Zambia, Africa.

The ramifications of this are potentially profound. Since earthquake activity has been observed on both Mars and the moon, we can seemingly infer the presence of underground water on these bodies. But here on Earth, the link between water and earthquakes may be an important first step in earthquake prediction.

Rock Dilatancy

Just before rocks fracture, they tend to swell or dilate. This phenomenon is called *dilatancy*. One theory is that the dilatancy is caused by tiny cracks that open up in the rock and fill with air. The overall volume of rock shows a slight but measurable increase as a result. P-waves passing through such rock experience a slower velocity because of the air pockets. P-waves travel slower than normal in rocks that are filled by air spaces, but are unaffected by spaces filled by water.

Gradually, ground water seeps into the cracks, displacing the air. The greater the volume of the cracks, the longer the time required for water to fill the cracked region. But when the cracks are finally filled by water, the time is ripe for an earthquake.

The theory is not only interesting, it provides a method for monitoring the process. As the rocks swell and air fills the cracks, the P-wave velocity slows. As water begins to fill the cracks, the P-wave velocity gradually rises, returning to normal. Thus, simple monitoring of P-wave speeds could, in theory, lead to earthquake prediction.

Electrical conductivity provides yet another means of monitoring the process. Dilated rocks with air filling the cracks can conduct electricity easier than can ordinary rock. Dilated rocks with water filling the cracks can conduct electricity easier still. So, as rocks dilate and their cracks fill with air and then water, a gradual increase in electrical conductivity can be seen.

Still a third measure is radioactive radon gas that is released, particularly from deep wells, prior to an earthquake. As ground water fills the cracks of weakened rock, much of it becomes contaminated by chemicals from the rocks. Radioactive radon is easy to monitor.

The record of the theory of rock dilatancy in predicting earthquakes is hopeful, but far from perfect. The most spectacular success was the successful prediction of a huge earthquake that hit Manchuria, China in February 1975. Perhaps tens of thousands of lives were saved as Chinese scientists predicted the correct day of the quake. Major centers of population were evacuated, and the estimated death toll was less than 500. Other examples of successful predictions exist, but none is so spectacular. The theory, unfortunately, has had its failures. The worst was the failure of the Chinese to predict the massive quake near the large city of Tangshan in July 1976. Although scientists were actively monitoring the clues according to the theory, the evidence was too ambiguous to make a firm prediction.

In 1979, a moderate 5.9 earthquake occurred south of San Francisco, near the town of Hollister, California. The quake occurred without any signs of swelling rocks, electrical changes, alterations in P-wave velocities, or radon emissions. Clearly, the theory of rock dilatancy is not the final answer.

Seismic Gaps

Another theory is based on the reasonable presumption that movement of a plate should occur at approximately equal speeds for all parts of the plate in a fault zone. If movement has occurred at points A and C, but not at B (which lies between A and C), then movement at B will soon occur. An adjunct to the theory is to observe foreshock behavior. An area may remain unusually quiet seismically for a long period. Then foreshocks may occur as a warning of an upcoming major quake.

This theory is eminently logical, and has met with some successes. But it lacks the amount of precision necessary to tell us the exact time of an earthquake down to within a few days.

Other Theories

Many other theories exist. The two mentioned above are the most firmly grounded in scientific principle. One theory, termed the *Cosmic Window Theory*, states that gravitational forces from the moon and planets, if aligned correctly, can disturb the tectonic plates enough to bring on an earthquake. The theory can be quickly refuted by a direct calculation of the force created by such an alignment, and a comparison with what force is required to bring on movement of tectonic plates. Although in theory it is possible for rocks to be at the instant of giving way to stress, and the planetary alignment to make the extra difference, the probability of the rocks lying in this special state is so minuscule that the theory can be safely dismissed.

Psychics have claimed to predict earthquakes correctly. Such claims must be critically analyzed, along with the percent failure rate. The ability of psychics to predict earthquakes can be dismissed on the basis of statistics.

CONCLUSIONS

Will it ever be possible to predict earthquakes as well as we can predict the weather? This question embodies the single goal of earthquake science. The implied comparison between earthquake phenomena and weather phenomena is worthy of thought.

Weather phenomena are created by masses of air of different types (e.g., continental, maritime tropical, maritime polar, etc.) that interact with each other. Convergent air masses create fronts where colder, denser air dives below warmer, lighter air, and precipitation results. When air masses diverge from each other, this allows low altitude air to rise through convection, leading to spotty thundershower activity. Finally, a stationary front can be formed when two air masses share a transform boundary, i.e., they slide past each other.

I have been a little loose with my terminology here in an effort to point out the similarities between the Earth's atmosphere and its mantle. The point is that the same kinds of phenomena that lead to instabilities in the atmosphere lead to instabilities of the mantle—large air masses or plates interacting with each other in various ways. In a way, this is not surprising. Temperature and density differences of air masses and tectonic plates are in both cases the governing factors in the interactions. From the standpoint of prediction, however, there are two big differences between the two phenomena.

The first difference has to do with instrumentation. We live in the air and breathe it. We fly airplanes through it. We can capture satellite images of it. The information is there for us to take.

We don't live in the Earth's crust, nor do we have any pressing need to go near it as part of our daily lives. It is no simple matter to measure properties of the Earth's interior with instruments. Often, we are limited to gathering data on the surface, looking at rock dilatancy or deciphering the complexities of seismic waves. From this kind of information, we must infer what is going on below the surface. It is like trying to deduce the size and shape of an elephant based on a single footprint found in the mud.

A second important difference between weather and earthquake prediction has to do with time scales. Air is lighter than rock, and it moves many times faster. Cold air dives under warm air in a day to form a front. An oceanic plate takes thousands of years to dive underneath a continental plate, and even then the process may continue.

Prediction of weather phenomena based on daily increments is reasonable, because it is a period of time consistent with the speed of the underlying processes.

Predictions of when earthquakes will occur as the Nazca Plate dives below the South American Plate deal with a different time scale. Trying to predict the day of the next earthquake is like trying to predict the time that it next starts to rain to the nearest five minutes.

Many hopes of predicting earthquakes have been based on trying to find the key cause-and-effect relationship. Every time it is cloudy, there is a threat of rain. What is the equivalent statement for earthquake prediction? Realistically, we probably cannot ever hope to understand the Earth's interior as well as we can understand its atmosphere. But finding a satisfactory answer to the above question is the point, for therein lies the nub of the mystery. If A does not happen, B will not happen. If A does happen, there is a good chance B will happen very soon. This kind of causal relationship exists between a cloudy sky and a threat of rain. A realistic goal is to find the same level of causal relationship for earthquake behavior.

A fertile area for exploration is in the analysis of seismic waves. This is where we find the best and most reliable information about the goings-on inside the Earth. The chances are reasonable that the key to unlocking the mystery lies in finding an imaginative new approach to analyzing the data and gleaning critical information from it. This is no easy task, however. We have already discussed the difficulties surrounding the wave equation of mathematics.

Another fertile area for exploration is to reexamine the underlying assumptions of earthquake analysis. As disconcerting as this sounds, it does need some thought. The theory of plate tectonics has been invaluable in helping us get a foothold into the infant field of earthquake prediction. But it does not explain everything. While the great majority of earthquakes occur at or near plate boundaries, many do not. One of the largest earthquakes on record was a massive jolt that struck southeast Missouri, in an area known as the "boot heel." The quake struck in 1811, and its magnitude has been estimated at a whopping 8.7. The Missouri boot heel lies smack in the middle of the North American Plate, a thousand miles from the nearest plate boundary. In the mid-eighteenth century, the British Isles were rocked by several moderate earthquakes. Again, the area was far removed from any plate boundaries. Quakes have also struck such unlikely areas as Boston, Massachusetts; Buffalo, New York; Charleston, South Carolina; and Montreal, Quebec, all long distances from plate boundaries.

Geologists have put forward explanations for these mysterious intraplate quakes. Friction between the plate and the asthenosphere is one explanation. Nonetheless, the intraplate quakes pose a challenge.

The theory of plate tectonics cannot explain them. Thus, the theory of plate tectonics is not the all-encompassing cohesive theory to explain earthquakes. It is a first step, indeed a massive first step, but it is not the last word.

We must also remember that the theory of plate tectonics is, in reality, just that—a theory. It is not proven fact. This model is widely accepted by the scientific community, but it does have dissenters.

One objection to the theory is that it offers no explanation for the forces responsible for moving the massive plates across the Earth. One possibility is convection currents, resulting from differences in temperature between sections of the Earth's crust. This is analogous to the Earth's atmosphere, where currents of warm air rise, then cold air rushes in to take its place, forming a pattern known as *convection.*

Another possibility is simple gravity drawing subducting plates downward toward the Earth's interior, and thus keeping the whole process moving. But this is arguable. Some of the plates are so thin, according to one counter-argument, that they would simply break apart if pulled by one end.

A major objection to the theory of plate tectonics arises from the complex behavior observed at some places where plate boundaries abut continents. The present plate boundaries in the areas of Greece, Yugoslavia, and Turkey are so complex that they cannot all be categorized as one of the basic three (divergent, convergent, or transform). To explain the observed goings-on in this region, a mass of *microcontinents* must be invented for the theory to hold up. The explanations that are thus constructed assume an artificial character and cast doubt on the whole theory. So we are left with a critical dilemma. On one hand, the theory of plate tectonics has taken us a long way, and it has become dear to our hearts. On the other hand, there is strong evidence to suggest the theory is incomplete. In science, no theory lasts forever. At some point, as we learn more, it is replaced by something better, something more complete.

It is possible that the key to unlocking the mystery of earthquake prediction lies in a more complete model to replace plate tectonics. Perhaps plate tectonics just needs some new additions or amendments. Or perhaps a creative new idea, one that is of comparable brilliance to Wegener's, is needed.

Meanwhile, the where and when of the next big quake assumes its rightful place as one of the most baffling mysteries faced by modern science today.

BIBLIOGRAPHY

Books

Bath, M. *Introduction to Seismology*. Basel, Switzerland: Birkhaser Verlag, 1979.

Bolt, B. *Earthquakes*. New York: W.H. Freeman and Co., 1988.

——. *Inside the Earth*. New York: W.H. Freeman and Co., 1982.

Bolt, B., ed. *Earthquakes and Volcanoes*. New York: W.H. Freeman and Co., 1980.

Eiby, G.A. *Earthquakes*. Auckland, New Zealand: Heineman, 1980.

Giancoli, D. *Physics*. 3rd ed. Englewood Cliffs, NJ: Prentice-Hall, 1991.

Hallam, A. *Earth Sciences: From Continental Drift to Plate Tectonics*. Oxford: Clarendon Press, 1973.

Karnik, V. *Seismicity of the European Area*. Dordrecht, Holland: Reidel, 1969.

Miller, R. *Continents in Collision*. Alexandria, VA: Time-Life Books, 1983.

Press, F., and Siever, R. *Earth*. 4th ed. New York: W.H. Freeman and Co., 1986.

Richter, C. *Elementary Seismology*. San Francisco: W.H. Freeman and Co., 1958.

Rikitake, T. *Earthquake Prediction*. Amsterdam: Elsevier Scientific Publishing Co., 1976.

Walker, B. *Earthquake*. Alexandria, VA: Time-Life Books, 1982.

Periodicals

Hamilton, R.M. "Quakes Along the Mississippi." *Natural History*, August 1980.

Herrman, Lisa. "Alien Worlds on the Ocean Floor." *Science Digest*, April 1981.

Hurley, P.M. "The Confirmation of Continental Drift." *Scientific American*, April 1968.

Mantura, Andrew. "The Mediterranean Enigma." *Oceans*, January 1977.

Nuttli, Otto. "The Mississippi Valley Earthquakes of 1811 and 1812." *Bulletin of the Seismological Society of America*, February 1973.

Powell, Corey. "Peering Inward." *Scientific American*, June 1991.

Can Volcanic Eruptions Be Predicted?

Volcanoes are among the most underrated of all natural phenomena. Most volcanoes lie in remote areas of the world, and their eruptions seem to have little effect on anyone. Much of the world regards volcanoes as objects of scientific curiosity and little more. The eruption of Mt. St. Helens in Washington State on May 18, 1980 alerted people somewhat to the power of volcanoes, finally giving these phenomena a measure of the respect they deserve.

The destructive power of the May 18th eruption can only be described as awesome. The mountain expelled almost 300 million tons of ash and rock. Sixty-six people were killed in an area as far away as sixteen miles. A scenic landscape was destroyed in minutes, and replaced with a bleak, ash-covered, lifeless terrain that looked more like the moon than our Earth. The U.S. Forest Service estimated that ten million trees, covering a 100-square-mile area, some over 100 feet tall, were destroyed instantly by the windstorm, their stumps reduced to splinters. The eruption rattled windows as far away as 100 miles. Ash fell on towns eighty or more miles away, covering the streets and rooftops and darkening the sky.

Yet, by some standards, the Mt. St. Helens eruption was not large at all. The worst volcanic tragedy of the twentieth century occurred on the island of Martinique in the Caribbean Sea in 1902. The eruption of Mt. Pelée killed almost all the 30,000 people living in the nearby town of St. Pierre. The second worst disaster of the century occurred in 1985, when the volcano Nevado del Ruiz erupted in a remote area of Colombia. Over 20,000 inhabitants of the town of Armero were killed by resulting mudslides. Armero was about thirty miles from the volcano.

The size of a volcanic eruption can be measured by estimating the volume of ejecta (ash, rock, and lava) expelled by the volcano during its eruption. In terms of this volume, the Mt. St. Helens eruption gave off about one cubic kilometer of ejecta. The eruption of Mt. Katmai in the Aleutian Islands in 1912 gave off about twelve cubic kilometers of ejecta. In 1883, Mt. Krakatoa in Indonesia gave off about eighteen cubic kilometers of ejecta. The blast from Krakatoa could be heard in Australia, thousands of miles away. A huge tidal wave was created by the explosion, killing several thousand people on nearby islands. Most of the island where Mt. Krakatoa stood was destroyed.

The largest eruption of modern times was Mt. Tambora, also in Indonesia, in the year 1815. Tambora was estimated to expel eighty cubic kilometers of ejecta, completely dwarfing the Mt. St. Helens blast. Besides the disastrous effect that this eruption had on human life, it had a noticeable effect on worldwide weather.

Temperatures around the world dropped an average of two degrees for an entire year. This does not seem like much, but some areas were affected more than others. In New England, for example, it snowed throughout the month of June and a killing frost in August destroyed many crops. Severe food shortages occurred in Europe because of widespread crop failures. The following year, 1816, is still referred to around the world as the "year without a summer." Low temperature records were set in that year that still have not been broken.

The modern world has so far been spared from an eruption of the size that occurred at Yellowstone about two million years ago. The Yellowstone volcano expelled an astounding 3,000 cubic kilometers of ejecta, making that eruption almost forty times the size of Tambora. An eruption of the same general magnitude as Yellowstone occurred on the Indonesian island of Sumatra about 20,000 years ago. Some scientists claim that the eruption was large enough to plunge the Earth into the last great ice age. Others say it was only a minor contributor to bringing on this ice age, and still others dismiss it entirely as having no significant effect. In view of the observed effects of Mt. Tambora on the world's weather, one must pause and wonder at what might be expected from an eruption forty times larger.

A massive earthquake can cost thousands of lives and create horrible hardships for those affected. What a massive volcanic eruption could do to modern society is anyone's guess. In the worst imaginable case, it could pose a serious threat to civilization as we know it.

But volcanoes are not all bad. In fact, volcanoes are a positive asset to the world. Many scientists believe that life would never have developed on Earth without the presence of volcanoes. Volcanoes are believed to be the major source of steam, carbon dioxide, and other chemicals used to make up the young Earth's oceans and atmosphere. Some scientists believe that the first life on Earth developed in the favorable temperature and pressure conditions afforded by volcanic hot springs.

Some evolutionary biologists believe that volcanoes are healthy to the development and advancement of living species. An environment that is too stable, and where life forms are not sufficiently challenged, may not provide the stimulus necessary for life to evolve to more advanced stages. On the other hand, an environment of catastrophic extremes could spell the end of life altogether. But the Earth, with its volcanoes and associated climate changes, strikes a rare and happy medium. Volcanic activity could stress life forms enough so that only those able to adapt to the changes would survive. In this way, the evolutionary process of natural selection would tend to raise the over-all survivability of living creatures. It is not far-fetched to say that mankind owes its existence to volcanic activity.

Volcanoes and volcanic hot springs offer us the tremendous potential of geothermal power. The heat within the Earth's interior is enormous, but it is difficult to reach because it is not concentrated. The exceptions are volcanoes and volcanic hot springs. The city of Reykjavik, Iceland is heated almost entirely by geothermal energy. The promise of geothermal energy is incredible. The U.S. Geological Survey has estimated that the total geothermal energy capability of the continental United States is equivalent to more than twice the world's oil reserves. The trick is to find a way to get at this vast resource easily and economically.

Volcanoes are responsible for some of the world's most fertile soils, probably because of potassium and phosphorus that are eventually absorbed into the ground from volcanic rocks and ash. Record crops of apples in Washington State were harvested in the years following the 1980 Mt. St. Helens eruption.

Volcanoes are responsible for the world's most prized gemstones. Precious rocks such as diamonds are forged deep in the Earth's interior under conditions of high temperature and pressure. Volcanoes then provide the mechanism for transporting them onto the Earth's surface. Volcanoes are also responsible for some of the most beautiful scenery in the world.

VOLCANO BASICS

Much can be learned about a volcano by doing nothing more than studying its shape. This is because a volcanic mountain is formed from material that the volcano erupts. Thus, general statements about the volcano's eruption history can be made based on the volcano's shape.

Volcano Shapes

The simplest type of volcanic shape is known as the cinder cone. This shape results from explosive eruptions that throw out mostly rocks. The rocks pile up in a steep hill around the vent. If lava flows out later, it can cement the loose rock into a cone shape.

The most famous and dramatic example of a cinder cone is Mt. Paricutin in Mexico. It was discovered in 1943 by a Mexican farmer, who noticed a large crack that had opened up in his land overnight.

Ash and hot stones were being spewed out of the crack. The astonished man was witnessing the birth of a volcano.

Mt. Paricutin grew at an astounding rate. It was thirty feet above the ground after one day, and grew another 500 feet its first week. It reached an eventual height of 1,353 feet and erupted almost continuously. Then, nine years later, in 1952, Paricutin suddenly fell silent, and has not erupted since.

Paricutin was typical of most cinder cone volcanoes, which grow very rapidly and die young. A cinder cone volcano can be recognized by its steep sides of loose rock. When cinder cones become inactive, rocks piled up around the cone fall in, plugging the vent. Then the cinder cone looks exactly like a normal mountain or hill, except for the loose rocks composing its sides.

Shield volcanoes are large, gently sloping domes. They are made by quietly flowing lava that erupts consistently and over a long period of time. The dome shape is much different from the steep-sided cinder cone. The term *shield volcano* describes the shape of this class of volcano, which resembles a warrior's shield turned upside down to lie on the ground. The dome of a shield volcano is thus not usually symmetrical, but is elongated in one direction. The volcanoes of Iceland and Hawaii are examples of shield volcanoes.

The composite volcano, or *stratovolcano*, is created by alternating cinder cone type eruptions with quiet lava flow eruptions. Most of the world's volcanoes fall into this third class. They can grow quite tall and have a long life span. Stratovolcanoes are considered the most scenic type of volcano. Examples include Mt. Fuji of Japan and Mt. Rainier of Washington State.

The eruptive history of a volcano determines its shape. Short-lived, violent eruptions of rock and lava give a cinder cone shape. Steady, quiet lava flow eruptions give a shield volcano. A mixture of the two types gives a stratovolcano.

A fuller picture would include a number of other factors that determine a volcano's shape. The volume of material expelled in a typical eruption is one such factor. Eruptions that involve small volumes of material tend to build up mounds close to the vent. Large volumes of erupted material create plateaus of lava or ash.

Therefore, the shape of the volcano in the area surrounding the vent provides some information about the volcano's recent eruptive history. A caldera, a crater with a much larger diameter than the volcano's vent, will often form following an eruption of a large ash plateau. If pressure from underlying lava becomes insufficient to support the volcano's cone, it can collapse inward on itself, forming a caldera. Crater Lake in Oregon is an example of a caldera. It was formed by the eruption of Mt. Mazama about 6,000 years ago. That massive eruption involved about forty-two cubic kilometers of ejecta.

The viscosity of the magma affects the explosiveness of eruptions, and is therefore another factor affecting a volcano's shape. Viscosity can be defined as a material's resistance to flow. For example, molasses has a higher viscosity than water.

Viscosity is important because it affects the growth of gas bubbles within the substance. If magma is very fluid (i.e., has a low viscosity), gas bubbles can form quickly and flow upward through the magma. In viscous magma, gas bubbles form, but face a large resistive force as they try to rise through the magma. The force of the gas bubbles on the viscous magma creates pressure that can explode violently in an eruption. Thus, more viscous magmas in general lead to more violent eruptions.

Eruptions of low viscosity, fluid magma form gently sloping shield volcanoes. Moderately viscous magma, causing moderately explosive eruptions, form stratovolcanoes. Highly viscous magma has the potential to create massive eruptions, creating ash or lava plateaus and possibly a caldera.

The geometry of the vent is another factor that determines a volcano's overall shape. In Iceland, volcanic vents are long cracks. An eruption along the length of the vent will form a lava flow pattern that is not symmetrical. If the vent is a crack moving north to south, most of the erupted lava will flow either east or west upon exiting the vent. The volcano's shape, over time, will assume the shape of a shield, with the shield's axis pointing north to south, paralleling the crack.

A volcano's shape is also determined by its location. A submarine volcano will be shaped differently from an above-ground volcano. Underwater volcanoes are always steeper than their counterparts on land. This is because water acts as a more effective cooling agent than land or air. Lava erupted from an undersea volcano will cool more quickly. It will harden before traveling as far. Water is also a more viscous medium than air, and thus resists the flow of magma through it more than does the air. For both these reasons, the result will be a volcano with steeper sides.

Volcanic Rocks

The two most abundant chemical elements in the Earth's crust are silicon and oxygen. Silicon dioxide, or silica, is a compound that consists of one part silicon and two parts oxygen. It is written SiO_2. The mass of the Earth's crust is 59% silica, and silica is a major component of 95% of all known rocks on Earth. Silica is an important ingredient in magma. All magma contains silica in varying amounts. The presence of silica in general increases magma viscosity.

The four major types of magmatic rocks, in increasing order of silica content, are basalt, andesite, dacite, and rhyolite. Basalt contains about 50% silica, while rhyolite contains about 75% silica. Of the four rocks, basalt is the most dense and rhyolite the least. As the silica content in these rocks increases, the viscosity increases and the density decreases.

Since the viscosity of the magma is an important factor in the potential explosiveness of a volcano's eruptions, magma consisting mostly of rhyolite is the most dangerous in terms of eruptive possibilities. Basaltic magma is common in the mild eruptions of shield volcanoes.

Types of Eruptions

The force that causes magma to rise through a volcano's vent is the same force that causes an iceberg to float on the ocean. As we discussed in the last chapter, this is the force of buoyancy. It is defined by Archimedes' Principle.

Magma deep in the Earth is less dense than the surrounding rocks that block its passage to the surface. The magma tries to rise to the surface, just as an ice cube tries to rise to the surface of a cup of water. The magma's path to the surface is blocked by overlying rocks, so the magma exerts pressure on the rocks as it tries to push its way up. Eventually, the pressure exerted by the magma is strong enough to break through the rocks. As the rocks give way, pressure on the underlying magma is suddenly released. It bursts out. An eruption!

As we have already discussed, the viscosity of the magma is a major factor in the explosiveness of eruptions. As magma becomes more viscous, its viscosity prevents gas bubbles from moving freely through it and eventually boiling away harmlessly. When gas bubbles become trapped in the viscous magma, the buoyancy force increases. This is because gas bubbles are less dense than magma, and a magma-gas combination is much less dense than magma by itself. Therefore, the more gas trapped in the magma, the greater the force pushing the magma upwards.

For similar reasons, the original density of the magma is a factor in determining explosive capability. Of the four major rock types that comprise magma, basalt is the most dense and rhyolite is the least dense. This means that magma made of rhyolite will not only be the most viscous but also the least dense. Thus, for two reasons, magma made chiefly of rhyolite has the most explosive potential.

The amount of gas available to be dissolved in the magma is also a factor. If more gas is available to be dissolved in viscous magma, the dissolving will make the magma more buoyant and increase the explosive potential. The rate at which pressure on the magma is lowered at the time of an eruption is a major factor. If magma exerts a large pressure on overlying rock, will it break through in one massive fracture of the rock, or will the rock hold, letting the magma escape more gradually through slowly opening vents?

The geometry of the vent is critical in determining the rate that the pressure is released. A small vent at the top of a cone-shaped volcano can be easily plugged by viscous magma. This can lead to a dangerous scenario with explosive potential, as the magma plugging the vent could suddenly yield to underlying pressure.

Besides the possibility of blockage, a long narrow passageway leading to a narrow exit is a dangerous vent geometry. Magma forcing its way from a spacious magma chamber into and through a long narrow tube travels at a very high speed through the tube. This phenomenon comes from a law in fluid mechanics called the Venturi Effect. (See Figure 3-1.) A simple way of stating this law is that the rate of flow of a fluid stays constant through a constricted channel.

We can see the Venturi Effect in our daily lives without even watching fluids. For example, if you are stuck in a traffic jam on the highway, it may be because of an obstruction ahead of you. Perhaps

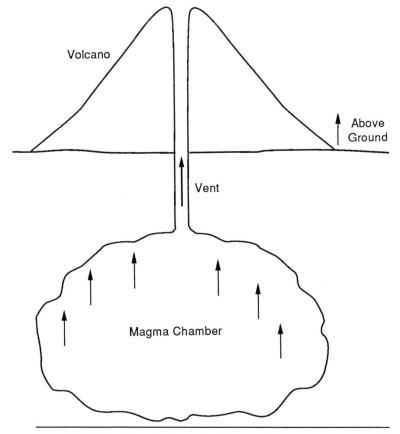

FIGURE 3-1. *The Venturi effect accelerates the movement of magma traveling from a spacious chamber through a narrow vent.*

one lane is available for traffic at the point of the obstruction, but three lanes are available to you now. If cars are passing the obstruction at anormal speed of thirty miles per hour, you are probably stuck in the traffic jam moving about one-third this speed, or ten miles per hour. The rate of flow of the traffic, in terms of cars per second, remains constant at all points. But where three cars must pass, they do so at one-third the speed of where one car must pass. A similar effect can be observed while waiting in line to enter a football stadium through a narrow gate.

The Venturi Effect can also be observed in fluids. A stream generally speeds up as it goes through a narrow section. Rapids in streams are therefore seen where the stream is narrowest. Winds are usually strongest as they pass through narrow canyons. This same principle can be applied to magma. Upward movement from a spacious chamber into a narrow passageway will accelerate it.

Still another factor affecting the potential explosiveness of an eruption is the temperature of the magma. Magma temperatures range from 1,300 degrees to about 2,200 degrees Fahrenheit. As the temperature of the magma increases, its contents become more liquid, and thus less viscous. "Cooler" magma is the most viscous, and thus the most capable of explosive potential.

The temperature of the magma also affects the type of ejecta hurled out during the eruption. High temperature magma of low viscosity will be mostly liquid. Lower temperature magma will contain increasing amounts of solid rock fragments. Volcanic ash, which is really not ash at all but minute rock particles, represents one form of rock that is expelled as ejecta. Other rocky ejecta include cinder-size fragments and blocks or "bombs," which can be larger than basketballs and quite dangerous. The 1968 eruption of Mt. Arenal in Costa Rica expelled a large number of destructive bombs that left the towns of Tabacon and Pueblo Nuevo littered with craters.

A particular danger of an eruption can occur if the hot gases expelled from the volcano are of low viscosity and heavier than air. In this case, they will not rise high into the atmosphere, but will rush down the sides of the volcano along the ground. Such a disaster occurred in the 1902 eruption of Mt. Pelée. Thousands were incinerated instantly by the rapidly moving hot gas.

WHAT MAKES LAVA SO HOT?

Lava is a product of the intense heat found in the Earth's interior. The temperature of the Earth increases with increasing depth below the surface at a rate of about seventy degrees Fahrenheit per mile. This tremendous rate of increase, called the *geothermal gradient*, is largest near the Earth's surface. As you near the Earth's center, the temperature still increases, but at a slower rate. Nonetheless, the figure gives some idea of the great amount of heat that lies within the Earth.

Some of this heat is left over from the infant Earth, when it was a mass of molten material that gradually cooled. Some scientists believe that a portion of the Earth's internal heat may come from a phenomenon called *tidal friction*. As the tides affect sea water, so they also affect rocks in the Earth's interior, moving them back and forth. As rocks move in this manner, they experience friction as they grind past each other, and thus generate heat.

No one knows exactly how much of the Earth's heat is left over from its early days, or how much is due to tidal friction, or how much is due to other sources. One definite source, however, that is responsible for perhaps most of the Earth's heat, is a process called *radioactive decay*, or simply radioactivity.

Radioactivity

In classical chemistry, each atom contains elementary particles called protons, neutrons, and electrons. Protons have a positive electric charge, and electrons have a negative electric charge. The negative charge of an electron is exactly enough to balance the positive charge of a proton.

An atom consists of a nucleus, which is made up of protons and neutrons together, and electrons, which circle the nucleus in orbits. According to the classical model, like electrical charges repel each other, and unlike electrical charges attract each other. Electrons in their orbits are not sucked into the nucleus, even though their charges are opposite. This is because they have momentum that keeps them in orbit, just as the Earth revolves around the sun without getting sucked into the sun by gravity. But what about all those protons in the nucleus? They should all be repelling each other since they all have the same electrical charge. Why doesn't the nucleus break apart?

To explain this, theoretical physicists invented the concept of *strong nuclear force*. It is stronger than the force of electrical attraction or repulsion, so it will hold an atom's nucleus together, but it only operates over a very small distance.

An atom must contain the same number of protons and electrons to be electrically neutral. The number of protons, which must exactly equal the number of electrons in such an atom, is called the *atomic number* and is unique for each chemical element. The atomic number of hydrogen, the lightest element, is one, that of helium is two, and so forth over the entire spectrum of chemical elements.

A chemical element is uniquely determined by the number of protons or electrons it has in its atoms. But it can have atoms with different numbers of neutrons. Nuclei that contain the same number of protons and electrons but different numbers of neutrons are called *isotopes*. An example is carbon. The atomic number of carbon is six, which means that each carbon atom has six protons and six electrons. But there are several different isotopes of carbon. Carbon-11 has five neutrons, carbon-12 has six, carbon-13, seven; carbon-14, eight; carbon-15, nine; and carbon-16, ten.

The number of protons plus the number of neutrons in an atom is called the *atomic weight*. Thus, the atomic weight of carbon-14 is fourteen, since it has six protons and eight neutrons. Isotopes are often distinguished by listing the element with its weight, such as carbon-14, or uranium-238.

The point is this. As more and more neutrons are added to an atom to create new isotopes, the nucleus of the atom grows in size. This affects the ability of the strong nuclear force to hold the nucleus together, since the strong nuclear force is most effective at very small distances. An electrical repulsive force, tending to separate protons in the nucleus, becomes stronger than the strong nuclear force, and the atom starts to disintegrate. This phenomenon is known as radioactivity.

Some common radioactive elements are uranium-238, carbon-14, and thorium-232. As the atoms of these elements disintegrate, they become other elements. For example, as uranium-238 breaks up, it becomes uranium-234, shedding neutrons in the process. Thorium-232 breaks up, shedding protons and electrons to become radium-228.

A radioactive process involves a net loss of mass. This sounds counter-intuitive (not agreeing with what we see in our world), but it is true, as laboratory research shows. For example, as thorium-232 breaks up to form radium-228, each thorium atom loses two protons, two electrons, and two neutrons, which, as it turns out, is one helium atom. So helium is a by-product of this reaction. Now the mass of one helium atom plus the mass of one radium-228 atom is less than the mass of one thorium-232 atom. It's strange but true. Somehow, as this happens, we wind up with less mass than what we started with.

You may have seen the famous equation $E = mc^2$ and wondered what it meant. The "E" stands for energy, the "m" stands for mass, and the "c" stands for the speed of light. It is an equation that describes how mass can be converted to energy. And this is what happens to our missing mass. It is converted into energy—heat energy in our case. This is why the Earth's interior is so hot. It contains a great many unstable elements undergoing radioactive decay. As radioactive decay occurs, heat is created.

VOLCANOES AND PLATE TECTONICS

Volcanoes are not everywhere. They are concentrated around the world in narrow bands. One such band is the so-called Ring of Fire that encircles the Pacific Ocean, which we discussed in Chapter 2.

Another band lies along the Mid-Atlantic Ridge, running through Iceland and the bottom of the Atlantic Ocean almost to Antarctica. Other bands of volcanoes confirm the pattern. Volcanoes and earthquakes are, to a large extent, co-located; that is, they happen in the same general parts of the world. These parts of the world, for the most part, lie at or near the boundaries of tectonic plates.

How Do Volcanoes Form?

Heat from the Earth's interior melts rock, creating magma. The magma is less dense than the solid rock surrounding it. Then, by Archimedes' Principle, the magma experiences a buoyancy force pushing it upward. To break through the surface of the Earth, it needs to find a weak spot in the Earth's crust. Magma may drift around for quite some time in the Earth's inner recesses before finally finding the weak spot.

Most such weak spots have already been found and mark the locations of existing volcanoes. But according to the theory of plate tectonics, the Earth's plates are constantly moving about, creating and relieving stress. This creates a dynamic situation in which weak spots are constantly being created and destroyed. Thus, some volcanoes grow old and die, while other new ones are created.

A very large class of weak spots exists at divergent and convergent plate boundaries. The vast majority of volcanoes exist along these zones. Volcanoes generally do not occur at transform boundaries.

Divergent Plate Boundaries

As two neighboring plates move apart from each other, a large weak area is created in the resulting hole. Magma wells up into the hole, creating a long string of volcanoes. Most volcanoes at divergent plate boundaries occur undersea. Along the Mid-Atlantic Ridge there is a long string of undersea volcanoes. The island of Iceland provides a unique spot on the Earth, where the Mid-Atlantic Ridge lies above the surface of the ocean. Typical volcanic activity on a divergent plate boundary can be seen by observing the volcanoes of Iceland.

Most Icelandic volcanoes are nothing more than open fissures. As you look at the fissures, you are seeing the holes between two oceanic plates pulling apart. The fissures are long and narrow. As magma flows out and cools, a volcano is formed that is shaped like a shield. The long axis of the volcano runs parallel to the fissure.

Oceanic plates are denser than continental plates. They ride lower on the asthenosphere than do the continental plates. In terms of the types of rocks, oceanic crust is normally composed of basalt, the most dense of the four rocks we have discussed. Basaltic magma is therefore ejected by the volcanoes of Iceland. Basaltic magma is the least viscous and therefore has the lowest potential for explosive eruptions. Since the fissure-type vents in Iceland have a much larger area than the small cones found on cinder cones or stratovolcanoes, the magma will not be accelerated through to the surface by the Venturi Effect. In general, volcanoes along such rift zones erupt quietly.

If the diverging plates are continental rather than oceanic, the situation is more complex. This is the case in the Great Rift Valley of East Africa. As the plates diverge, the land mass left between the plates collapses, forming the valley. Magma rises through the resulting fractures, creating volcanoes.

Volcanoes that form in the Rift Valley of Africa are quite different from the volcanoes that form in Iceland, despite the fact that both groups arise from the middle of diverging plates. The African volcanoes, such as Mt. Kilimanjaro and Mt. Kenya, form the characteristic shape of stratovolcanoes, reaching tall heights and topped by roundish cones. They are quite different from the shield volcanoes of Iceland, which have the shape of rounded domes.

One important difference is in the type of rock forming the two different plates. While an oceanic plate is made up mainly of dense basalt, a continental plate is made up of a mixture of basalt, andesite, dacite, and rhyolite. The material of the continental plate is thus less dense and more viscous than that of the oceanic plate. Since this material makes up the magma that feeds the volcanoes, the volcanoes of the Rift Valley in Africa tend to erupt more explosively and with more viscous material. The shape of the volcano, which is heavily influenced by the eruption pattern, will therefore be different.

Convergent Plate Boundaries

The most common and explosive stratovolcanoes form at convergent plate boundaries. The type of volcanic activity that results from converging plates depends on the types of plates involved.

Volcanoes are rare in a region where two continental plates converge. In this case, subduction does not occur. The two plates smash into each other head-on, and folding results. Mountains are a direct result of this process. Volcanoes sometimes form, but, in general, the folding process does nothing to create magma.

When both plates involved are oceanic, one plate subducts, or dives under the other. The heavier, denser plate dives below the lighter plate. As the heavier plate dives into the Earth, eventually the Earth's heat becomes sufficient to melt the rock in the plate. Magma is thus created from the basaltic rock in the subducting plate. As the magma thus formed rises through the crust of the other plate, it is usually hot enough to melt some of the rock in the lighter plate. The lighter plate's rocks will usually contain the less dense rocks—andesite, dacite, and rhyolite. The magma created will be from a mix of rock types, and will be viscous magma.

Although the entire process is not completely understood, it is believed that a weak spot is created in the upper plate as it buckles from the contact with the heavier, subducting plate. The magma "finds" the weak spot and breaks through the surface of the Earth to form a volcano. An ocean trench marks the spot where the heavier plate dives below the lighter plate. A volcanic island arc is formed behind the trench.

The depth the subducting plate must reach in order for magma to form is usually about sixty miles. Magma will rise up directly from this point and will usually find an associated weak spot at which to form the volcano. The distance from the trench to the island arc is usually about 100 miles, roughly speaking. This distance can be approximated by geometry, assuming the angle that the subducting plate makes is about thirty degrees. (See Figure 3-2.)

Island arc volcanoes are responsible for some of the world's most spectacular scenery. The Aleutians, Philippines, Kuriles, islands of the Caribbean, and island chain around New Zealand are all examples of this type of volcano building.

The third type of converging plates occurs when one of the plates is oceanic and the other is continental. In this case, the oceanic plate, since it is always the denser of the two, subducts beneath the continental plate. The process of magma formation is exactly the same. An ocean trench is formed at the point of subduction. At a distance of

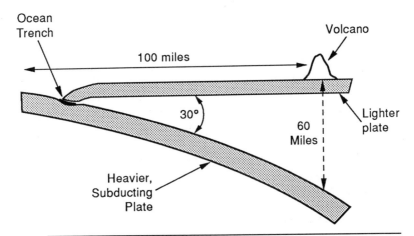

FIGURE 3-2. *The geometry of island arc creation. Assume a typical subduction angle of thirty degrees and a vertical depth of sixty miles where magma forms. The distance from the trench to the island arc can be estimated at about 100 miles, from the geometry of the 30-60-90 triangle.*

perhaps 100 miles inland from the trench, a volcano is formed over the continental plate on the land mass. The Pacific Northwest contains a string of several volcanoes that have been formed in this way. The entire west coast of South America contains a similar chain.

Subduction volcanoes are the most common on Earth. They are also the most dangerous, because they erupt magma that can be extremely viscous. They are also high and cone-shaped. This leads to the possibility of greater explosiveness because of the Venturi Effect. It also makes possible the dangerous situation in which the small vent can become plugged by hardening magma. As pressure builds up on this type of blockage, an explosive eruption can result if the blockage gives way suddenly.

Intraplate Volcanoes

Some volcanoes form in the interior of plates, a long distance from any plate boundary. Some of these volcanoes are the result of random weak spots that occur in the interiors of plates. But a second interesting class is the so-called "hot spot" volcanoes.

These volcanoes form from a persistent magma source that remains in the same spot over an extended period of time. A plate passes over the stationary hot spot. The plate moves, but the hot spot remains stationary. The result is a chain of volcanoes in the interior of a plate. The most famous example of this phenomenon is the Hawaiian Islands. This island chain is in the middle of the Pacific Plate. The newest volcanoes are on the island of Hawaii, at the southeast end of the chain. As you proceed from the island of Hawaii northwestward, you encounter islands that are progressively older, all formed by volcanoes.

The explanation of this phenomenon is that a stationary hot spot exists below the Pacific Plate. The hot spot is currently in the vicinity of the island of Hawaii. Over the years, the Pacific Plate has moved in a northwest direction. This movement has resulted in the creation of the string of volcanoes and hence the island chain, from northwest (oldest) to southeast (youngest).

Hot spot volcanoes are normally non-explosive, probably because their magma is mainly basaltic, with a low viscosity. While the Hawaiian Islands are the most famous example of a string of hot spot volcanoes, the Azores and the Galapagos Island chains are other examples of this interesting occurrence.

Volcanoes in the Solar System

The Earth is certainly not the only world on which volcanic activity exists. Volcanoes, in fact, seem to be a rather common development on many worlds. The most spectacular extraterrestrial volcanic observation was made in 1979 as the *Voyager I* spacecraft flew by Io, one of the many moons of Jupiter. Several volcanoes were seen in the act of erupting. Volcanic activity on Io is so great that scientists attribute the multi-colored appearance of this world to volcanic activity.

The largest volcano ever observed is the gigantic Olympus Mons on the planet Mars. It is about fifteen miles (about 80,000 feet) tall. Its roughly circular base is almost 400 miles in diameter. Scientists theorize that Olympus Mons may be a hot spot volcano. Unlike the Earth, however, with moving plates, the Martian hot spot continued to feed magma to the same stationary spot, causing it to grow over millennia into such a single enormous shape.

The observation of Olympus Mons and the theorized explanation of this huge volcano pose an interesting mini-mystery about the red planet. Since seismic activity has also been observed on Mars, does this make plate tectonics a possibility on Mars? And yet, how does this notion mesh with the theorized explanation of the enormity of Olympus Mons, which says that plate tectonics must not exist on Mars? Could the earthquake activity on Mars be caused by the same phenomena that produce the Earth's intraplate quakes? What could be going on to stress the rocks of Mars and produce quakes if it is not some kind of plate movement somewhere? The questions go on and on.

The large basins on our moon are believed to have been caused by volcanic activity in its past history. For years it was thought that the circular craters of the moon were caused by volcanoes, and were actually volcanic caldera. But the current meteor impact theory eventually won out, once it was understood that the impact of a meteor of any shape would produce an approximately circular crater. Nonetheless, the large basins, or *maria*, on the moon can seemingly be explained only by some past volcanic activity.

Examination of the surface of Mercury also shows signs of volcanic activity. Of particular interest is an area known as the Caloris Basin, an unusually smooth area on the otherwise heavily cratered planet. The most logical explanation for this region seems to be volcanism.

Scientists also believe volcanoes exist on Venus, but this case is not so clear-cut. Some evidence comes from observing the gas sulfur dioxide, which is erupted by most volcanoes on Earth. Rapidly changing fluctuations of sulfur dioxide in Venus's atmosphere have caused scientists to speculate that volcanic eruptions have been occurring there. Further evidence of volcanic activity on Venus comes from a general analysis of its atmospheric makeup. Scientists observe that the gases occurring in Venus's atmosphere are the same that are erupted from volcanoes.

VOLCANOES AND WEATHER

Earlier in this chapter, we discussed how the 1815 eruption of Mt. Tambora caused a noticeable climatic cooling. Most scientists today agree that volcanic eruptions can exert an effect that cools the climate, although they disagree as to the degree.

A current topical discussion concerning the so-called "greenhouse effect" says that certain emissions into the atmosphere will create a warming effect. How can this be? Why do emissions from a volcano cool the climate, but emissions from manmade processes warm the climate? What is going on here? To understand this interesting subject, we will take a brief excursion into the physics of heat.

What is Heat?

An amazing result of theoretical physics, which is confirmed by experiments, tells us that heat is nothing more than atoms or molecules in motion. As molecules move faster, they heat up.

This does not mean that windy days are warmer than calm days because air molecules are moving faster. There is a subtle difference. Differences in pressure generate wind, and air molecules are swept along from the region of high pressure to the region of low pressure. They are not moving because of their own energy. They "go along for the ride." If you boil water, on the other hand, you supply energy to the water molecules. Because of their higher energy, they move faster. They heat up.

It is important to discuss heat as moving particles, because moving particles generate waves, as we discussed in Chapter 2. Molecules, atoms, protons, and electrons all experience a displacement from their equilibrium positions as they are made to move faster. They vibrate back and forth in simple harmonic motion, generating waves.

Heat and Waves

The two properties that describe every wave are its frequency and its wavelength. If we picture a wave as a sequence of peaks and troughs moving past us, the frequency of the wave is the number of peaks that pass by per second. Frequency is usually expressed as cycles per second (the same thing as peaks per second), or Hertz. The wavelength of a wave is the distance between peaks. The speed of the wave is then the frequency times the wavelength. Thus, if a certain wave has a wavelength of two feet, and its frequency is ten Hertz, this means that ten peaks are passing by each second and the distance between each peak is two feet. The wave's speed is then twenty feet per second.

In terms of heat, hotter temperatures produce waves of higher frequency. This is easy to see conceptually. As particles vibrate back and forth faster because of a higher temperature, they will create waves of more cycles per second. The frequency of the waves produced equals the frequency at which the particles are vibrating. We also say that the energy of a wave increases with its frequency. This is natural, since the frequency depends on the speed of vibration of the heated particles.

Some particles vibrating back and forth are protons and electrons. These particles carry an electric charge. When such electrically charged particles move back and forth in simple harmonic motion, the result is a special type of wave known as an *electromagnetic wave*. If we had sensitive enough instruments, we could detect a changing electric field as such an electromagnetic wave passed by. The waves that reach us from the sun and the waves that the Earth radiates back to space are both examples of electromagnetic waves. In fact, all forms of heat generate electromagnetic waves.

There are three methods of transferring heat. Conduction involves direct contact with a hot object. Convection involves the mass movement of heated molecules from one location to another, and is important in meteorology. Radiation is the transferring of heat by means of electromagnetic waves. This is how heat is transferred from the sun to the Earth, and from the Earth to space.

Radiation from the Sun to the Earth

The light we see is electromagnetic radiation that our eyes have the ability to detect. Our eyes can detect radiation that is in a certain frequency range. Frequencies in the range of about 4×10^{14} Hertz to about 7×10^{14} Hertz are detectable as light by our eyes. The light at the low end of this frequency range is red. The light at the high end is violet. Other colors are spread throughout this range of frequencies.

Violet light has a higher frequency, and therefore more energy, than red light. But since all light travels at the same speed, red light has a higher wavelength than violet light.

About half of the electromagnetic radiation that reaches the Earth from the sun is in the visible range. The rest has either higher or lower frequencies than visible light. The radiation with higher frequency is called *ultraviolet*, and the radiation with lower frequency is called *infrared*.

Radiation from the Earth to Space

Heat that is stored in the Earth escapes off into space in the form of electromagnetic radiation. Of course, the Earth is not nearly as hot as the sun, so the radiation given off by the Earth will have a lower energy than that given off by the sun. Electromagnetic waves coming from the Earth will therefore have a lower frequency than the waves coming from the sun. The electromagnetic waves radiated from the Earth's heat fall in the infrared category. They have smaller frequencies but longer wavelengths than radiation in the visible range.

The Earth's Heat Budget

For the Earth to remain at a constant average temperature over time, it must have a balanced heat budget. The amount of heat coming to the Earth must equal the amount leaving the Earth to go off into space. Of course, such a situation is an idealization and never happens in reality. Earth's average temperature is always slowly changing over time. But sudden, dramatic shifts in the heat budget should cause us concern.

Of the radiation that reaches Earth from the sun, about 26% is immediately reflected by clouds. Another 7% reaches Earth but is immediately reflected. This 7% is determined by the topography. Ice reflects more radiation than grass. An empty field reflects more radiation than a forest. In general, an object's ability to reflect depends on its perceived brightness.

The Earth's *albedo*, or its ability to reflect, is estimated at 33%, and is the sum of 26% (reflection by clouds) and 7% (reflection by land). Of the 67% that is not reflected, 45% is absorbed by the ground, and 22% is absorbed by the atmosphere. All the heat absorbed by the Earth is eventually radiated out. Some escapes into space (about 79%), and the rest is absorbed by the atmosphere.

The key variable in the heat budget is the atmosphere. The atmosphere gets some of its heat from the sun and some from the Earth. What does it do with its heat? Like the Earth, most of the heat in the atmosphere is radiated back into space. Some, however, is recycled back to Earth. The Earth then radiates this heat, and so forth.

This sounds complicated, and it is. Earth, the atmosphere, the sun, and space all exchange heat with each other. (See Figure 3-3.) But the relationship between the Earth and the atmosphere is unique, since there is "feedback" or an exchange between the two of them. Some of what the atmosphere gives to the Earth, the Earth gives back, and vice versa.

There are many factors that affect the Earth's heat budget. Among them are:

1. The amount of radiation emitted by the sun.

2. The amount of solar radiation reflected or absorbed by the atmosphere.

3. The amount of solar radiation reflected by the Earth.

4. The amount of radiation from the Earth absorbed or reflected by the atmosphere.

In the above classification, we can do nothing about point (1). Point (3) depends on topography. It is certainly affected by the mass destruction of rain forests, but is beyond the scope of our discussion here. Points (2) and (4) are key to our discussion. Volcanic ejecta have

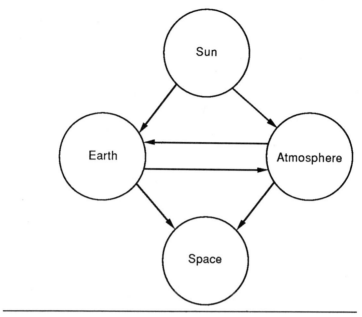

FIGURE 3-3. *The interacting entities in Earth's heat budget.*

a direct effect on (2). Manmade pollutants have a direct effect on (4).

Absorption and Wavelength

Imagine a group of logs lying end to end in a lake. Waves from the lake run up to the logs. Do the logs stop the waves? Or do the waves simply pass through the logs, lifting the logs up and down as they pass?

The answer depends on the wavelength of the waves. If the wavelength is very long, the waves will pass through the logs. If the wavelength is very short, the logs will stop them. Short, choppy waves will not make it past the logs. Long, undulating waves will. The boundary point is where the wavelength of the wave equals the width of the log. If the wavelength is longer, it can pass right through. If it is shorter, the log will stop it, absorbing some of its energy and reflecting the rest back into the lake. It is because of this principle that volcanic ejecta can cause a cooling effect.

Strong volcanic eruptions can propel gases far above the lower atmosphere into a region called the *stratosphere.* Particles injected into the stratosphere tend to stay there for a long time. There are two reasons.

First, the stratosphere lies above the precipitation producing region of the atmosphere, called the troposphere. Whereas material injected into the troposphere is normally removed from the atmosphere by precipitation, material injected into the stratosphere cannot be precipitated.

Second, the temperature in the stratosphere increases with altitude, unlike the troposphere. This fact makes vertical movement very difficult in the stratosphere. Air that is warmer than its surroundings will rise. Air that is cooler will sink. If a mass of air in the stratosphere starts to sink, it will be warmer than the air below it, and eventually rise again. Therefore, particles injected into the stratosphere are usually present for a long time.

A particular gas, sulfur dioxide (SO_2), is emitted from volcanoes during eruptions. If the eruption is strong enough to launch this gas into the stratosphere, a troubling condition occurs. The gas combines with available oxygen and water vapor to form an aerosol of sulfuric acid (H_2SO_4). An aerosol is a particle that is suspended in the atmosphere. Aerosols come in various sizes, but this particular aerosol contains particles whose diameters are about the same length as the wavelengths of visible light.

The situation is like that of the logs on the lake. The radiation from Earth, which is in the infrared range, will have a wavelength that is longer than the size of the particles. So it will simply pass through them. Of the radiation coming from the sun, much of it will appear as short, choppy waves, and will be blocked. Some will be reflected, and some will be absorbed. The effect will warm the stratosphere but cool the Earth. Earth will continue to radiate heat but will receive less. Since particles in the stratosphere tend to remain suspended there for quite some time, the condition could persist for several years.

Absorption and Energy States

Earlier in this chapter, we discussed the classical model of the atom, and how its nucleus of positively charged protons should split apart. The construct of the strong nuclear force was created to prevent this from happening. But a further problem exists.

It has to do with the orbiting electrons. Electrons are oppositely charged from protons, and hence are attracted to the atom's nucleus. Their momentum keeps them from falling in. The situation is similar to Earth revolving around the sun. Eventually, the electron, just as the Earth at some time, must lose energy to friction. Its orbit will decay, and eventually the electron will crash into the nucleus. Why doesn't this happen?

The riddle was solved with the invention of quantum mechanics, a modern and difficult branch of physics. The main idea of quantum mechanics is that atoms can exist only in certain energy states. An atom can have one unit of energy, or two units, and so forth. But it can't have one and one-half or two and one-eighth.

This idea is counter-intuitive—it does not agree with what we see in our world. When you are driving your car and take your foot off the gas pedal, you see the car gradually slow down. It doesn't happen that one second you're doing thirty miles per hour, and the next second you're doing twenty. Quantum mechanics says that there really are these sudden changes going on, but you don't notice them because they're closely bunched. The effect becomes drastic only at the atomic level, where electrons change energy states quite dramatically.

One success of quantum mechanics has been the explanation of spectroscopy. If an atom is bombarded with radiation, it will absorb some of the radiation and convert the absorbed radiation into energy. But because of the rules of quantum mechanics, it will only absorb the bits of radiation that will allow it to move to another acceptable energy state.

If a certain material is bombarded by light rays, which are then passed through a prism to separate the result into component colors, certain dark lines will appear. The dark lines correspond to the radiation that was absorbed by the material. The energy of light is proportional to its frequency, and only certain energy changes are acceptable. This means that only the light of corresponding frequencies of light is absorbed, creating the dark lines.

Each substance also has an emission spectrum. This is radiation given off by the substance as it moves to a lower energy state. It appears as a series of bright lines. Figure 3-4 sketches the basic apparatus for observing a substance's spectral lines. Interestingly, the emission spectrum of any substance is exactly the same as its absorption spectrum. This makes sense only by considering the laws of quantum mechanics. Only certain transitions to certain energy states are allowed. When the substance absorbs radiation, it makes a transition to a higher energy state. When it emits energy, it makes a transition to a lower energy state. Whether absorbing or emitting, the differences of the energy states are always the same. These differences

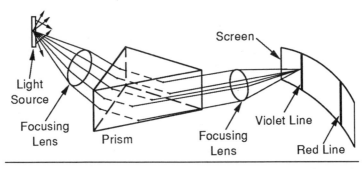

FIGURE 3-4. *Basic apparatus to observe a substance's spectral lines.*

in energy states correspond to the lines in the substance's spectrum.

Carbon dioxide and water vapor each have absorption spectra that include large portions of frequencies in the infrared range. There is no particular reason for this. It is just chance, and the fact that the acceptable energy states for these substances have the appropriate differences. This means that they absorb large chunks of infrared radiation to make transitions to higher energy states.

As a result, much of the infrared radiation given off by Earth never makes it to outer space, but is absorbed by these gases. That is fortunate; otherwise, the Earth would be about seventy degrees Fahrenheit colder than it is today.

As man-related processes pump more of these gases into the atmosphere, the heat budget of the Earth is disturbed. More than the usual amount of infrared radiation is captured by the atmosphere, and returned again to Earth. The result is a gradual warming of the Earth. This is the so-called "greenhouse effect."

THEORIES OF PREDICTION

Have you ever listened to the steady drip-drip-drip of a leaking faucet? You can almost guess the time of the next drip. They come at regular intervals. Why? Figure 3-5 illustrates the sequence of steps that go into the formation of a water droplet. As the diagram suggests, there is an attractive force between the water and the tube or faucet that holds it.

In fact, there is a weak attraction existing between all atoms. This attraction, called the van der Waals forces, is distinct from chemical binding and considerably weaker. (Chemical bonds involve electrons shared between atoms.)

The van der Waals forces are electric in nature. As electrons orbit a nucleus, the positive charge of the atom is centered in the nucleus,

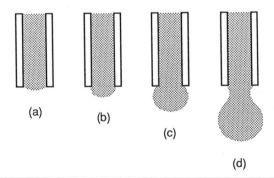

FIGURE 3-5. *Different stages in the formation of a water droplet.*

but the negative charges (electrons) are always moving. In principle, we can imagine taking a snapshot of the atom at one instant in time and plotting the locations of all the electrons. We could calculate the "average location" of the electrons, or the "center of mass" as it is called. This center of mass will lie in empty space, not coinciding with any one electron. (See Figure 3-6.)

The center of mass of the positive charges will not coincide with the center of mass of the negative charges, unless we are extremely lucky when we take our snapshot. Therefore, an electric field is set up. Neighboring atoms, which have their own electric fields, tend to orient themselves accordingly with our electric field. The result is a slight attraction, called the van der Waals forces, that exists between all atoms. Were it not for the van der Waals forces, a leaky faucet would run continuously, not drip at regular intervals.

Enough water must accumulate in a droplet so that the gravity on the droplet is sufficient to overcome the van der Waals forces attracting the droplet to the end of the faucet. If water accumulates at the end

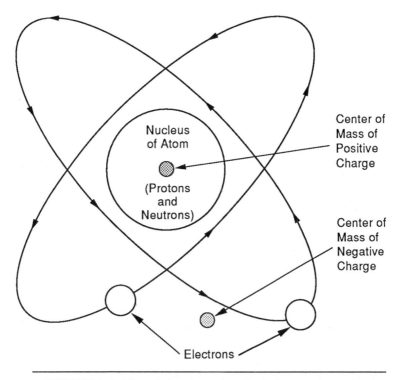

FIGURE 3-6. *The origin of van der Waals forces. At any instant of time, the centers of mass of positive and negative charges do not coincide, and a miniature electric field is set up.*

of the faucet at a steady rate, the drips will occur at equal intervals.

Eruptive History

You may have already made the analogy of the leaking faucet to the erupting volcano. A weak spot in the Earth's crust is like the bad plumbing. Magma is like the water. The volcano's vent is like the faucet. The difference is that it is not gravity, but buoyancy of the magma that eventually becomes great enough to overcome the van der Waals forces between the magma and rock surrounding it.

Of course, after reading this chapter, you know that there is much more to it. Pressure has to build up to overcome blockages, for example. But, in its simple form, the theory is that some volcanic eruptions can be predicted by carefully studying the eruptive history.

Eruptive materials from volcanoes are laid down on the ground, one on top of the other, in rows or *strata*. By searching through the strata, scientists can apply techniques to find the approximate times that the rocks were laid down, and thus estimate the time intervals between eruptions. Magnitudes of the eruptions can be estimated by the amount of rocky material in each strata and the extent of its distribution geographically.

Some volcanoes show consistent behavior, and these are the easiest to predict. Others are erratic. The 1980 eruption of Mt. St. Helens was roughly predicted by scientists at the U.S. Geological Survey, who, in 1978, stated that the volcano was likely to erupt explosively before the end of the century. Their prediction was based on the eruptive history of the volcano. They found that, over the last 4,500 years, eruptions had occurred about once every 225 years, and showed some signs of regularity.

Magma Watching

Every volcanic eruption is preceded by the movement of magma. Monitoring the movement of the magma underground can give important clues about an imminent eruption. There are several possible methods for doing this. Watching ground deformation along the volcano's sides or in the volcano's general vicinity is one method. As magma attempts to rise to the surface, it creates added pressure on nearby rocks. Ground uplift can usually be observed. A sensitive device called a tiltmeter detects such spots. The general pattern of ground uplift over time in the vicinity of the volcano can be used to trace the movement of magma masses underground.

Electrical soundings provide a second possibility. Magma is a much better electrical conductor than solid rock. This means that an electric pulse can travel more easily through magma than solid rock. The idea, then, is to take a series of electrical measurements around the volcano and in this way calculate where the magma must be located and how it is moving.

Small earthquake swarms in the vicinity of a volcano are a near-certain sign that underground magma is moving. Such earthquakes sometimes signal an impending eruption.

CONCLUSIONS

Can volcanic eruptions be predicted as well as the weather? It was a similar question that I posed earlier with regards to earthquake prediction. Volcanoes and earthquakes are related phenomena. It is reasonable that the same question be asked.

The answer to the question is similar in both cases. To predict volcanic eruptions as well as we can predict the weather, we must be able to monitor the movements of underground magma with the same accuracy that we can monitor above-ground air movement. The development of new technology to probe underground activity is a fertile area for research.

The analysis of the eruptive histories of volcanoes is also an area that promises to reap benefits. At first thought, it seems that this sort of technique lacks the precision needed to predict an eruption to within a few days. But new ideas are likely to emerge from a careful integration of the data collected on eruptive histories of many volcanoes around the world. The problem is that such work is enormously time-consuming.

The theory of plate tectonics, which had much to say about earthquakes, also plays a role in volcanology. We must remember that this theory is not proven fact, and there is much to suggest that it is incomplete. As in the case of earthquake prediction, the final answer to the volcano problem may involve a new scientific model to replace the cherished theory of plate tectonics. The development of such a model, which would have to be consistent with all the available facts as we know them today, would have to be nothing short of an act of superhuman creative wizardry.

But therein may well lie the solution to the mystery. Volcanoes are awesome things. Many scientists credit them with the development of life on Earth. Some also fear that a massive volcanic eruption in modern times could spell the end of much of that life. Being able to predict volcanic eruptions may only be the first part of the two-part question. If we can predict that a big one is coming, what can we do about it?

BIBLIOGRAPHY

Books

Cairns-Smith, A.G. *Seven Clues to the Origin of Life*. Cambridge: Cambridge University Press, 1985.

Cas, R.A.F., and Wright, J.V. *Volcanic Successions*. London: Allen & Unwin, 1987.

Decker, R., and Decker, B. *Mountains of Fire*. Cambridge: Cambridge University Press, 1991.

Editors of Time-Life Books. *Volcano*. Alexandria, VA: Time-Life Books, 1983.

Fisher, R., and Schmincke, H. *Pyroclastic Rocks*. Berlin: Springer-Verlag, 1984.

Francis, P. *Volcanoes*. Middlesex, England: Penguin Books, 1976.

Giancoli, D. *Physics*. 3rd ed. Englewood Cliffs, NJ: Prentice-Hall, 1980.

Harte, J. *Consider a Spherical Cow: A Course in Environmental Problem Solving*. Mill Valley, CA: University Science Books, 1988.

Holland, H.D. *The Chemical Evolution of the Atmosphere and Oceans*. Princeton, NJ: Princeton University Press, 1984.

Krafft, M., and Krafft, K. *Volcanoes: Earth's Awakening*. Maplewood, NJ: Hammond, 1980.

Marcus, R. *Volcanoes and Earthquakes*. Revised ed. New York: Franklin Watts, Inc., 1972.

Miller, R. *Continents in Collision*. Alexandria, VA: Time-Life Books, 1985.

O'Neil, P. *Gemstones*. Alexandria, VA: Time-Life Books, 1984.

Poynter, M. *Volcanoes*. New York: Julian Messner, 1980.

Stommel, H., and Stommel, E. *Volcano Weather*. Newport, RI: Seven Seas Press, 1983.

Wyllie, P. *The Way the Earth Works*. New York, Wiley, 1976.

Periodicals

Bullard, F. "Studies on Paricutin Volcano, Michoacan, Mexico." *Bulletin of the Geological Society of America*, May 1947.

Cook, R.J., et al. "Impact on Agriculture of the Mount St. Helens Eruptions." *Science*, January 2, 1981.

Decker, R., and Decker, B. "The Eruptions of Mount St. Helens." *Scientific American*, March 1981.

Findley, R. "Mount St. Helens: Mountain with a Death Wish." *National Geographic,* January 1981.

Fryer, P. "Mud Volcanoes of the Marianas." *Scientific American*, February 1992.

Hoblitt, R.P., et al. "Mount St. Helens Eruptive Behavior during the Past 1,500 Years." *Geology*, November 1980.

Sigurdsson, H., and Sparks, S. "An Active Submarine Volcano." *Natural History*, October 1979.

Stommel, H., and Stommel, E. "The Year Without a Summer." *Scientific American*, June 1979.

What Killed the Dinosaurs?

The dinosaurs roamed the Earth for a period of about 140 million years. They first appeared about 210 million years ago, and disappeared about sixty-five million years ago. During most of this vast time interval, they were rulers of the Earth.

To put these numbers in perspective, the oldest fossils of man show that he probably appeared on the Earth about two million years ago. During most of this time, however, man could not be considered ruler of the Earth. Man's rise to ascendancy was painfully slow. He spent much of his history scraping out a meager existence by gathering berries and cowering in fear from the wild beasts. It has been for perhaps 10,000 years that man can truly say he has been the ruler of the Earth.

In other words, dinosaurs ruled the Earth for a length of time equal to 14,000 times as long as man has ruled it. On a slightly different scale, if dinosaurs ruled the Earth for one day, then man has ruled the Earth for six seconds. The dinosaurs were eminently successful. When all is said and done, history may someday prove them to be more successful than man.

Most people today consider dinosaurs dead and gone, but this statement is only approximately true. Descendants of the dinosaurs live on, even today. Study of the fossil records shows striking similarities in the skeletons of modern birds to the ancient dinosaur Archaeopteryx. Some ornithologists classify all living birds as dinosaurs. Although this classification is disputed by most, almost all agree that birds originated from dinosaurs and are their descendants.

The characteristics of the crocodile's skull and the placement of his teeth in sockets give the crocodile two major characteristics of dinosaurs. Crocodiles are quite definitely a living link to dinosaurs. Whether or not we call them dinosaurs seems mostly a matter of word choices.

Despite the birds and the crocodiles, no one can deny that a dramatic event occurred some sixty-five million years ago that wrenched control of the world away from the dinosaurs. That event paved the way for the mammals, and thus for man. We unquestionably owe our existence on the Earth today to whatever killed the dinosaurs. It is for this reason that the mystery holds so much fascination for us.

HOW FOSSILS ARE FORMED

When any living organism dies, a process of decay begins at once. The decay results mostly from aerobic bacteria (i.e., bacteria living only in the presence of oxygen). Besides this type of decay, the parts of an organism can also be dissolved by water if given enough time. If left unprotected, the body of the organism also faces the danger of being destroyed by scavengers.

For these reasons, the creation of a fossil is a rare event. Rapid burial of the organism must occur quickly in order to protect it from scavengers, water, and aerobic bacteria.

Once an organism is entombed in this way, a process known as *petrifaction* (turning to stone) can occur on the organism's hard, bony parts. Minerals such as calcite, silica, or iron seep into the pores of the bony material, changing its chemical composition but maintaining and preserving its original shape.

Soft parts of animals or plants are less resistant to decay than hard parts, and thus are almost never preserved. The frozen remains of ice-age mammoths are noteworthy but extremely exceptional cases.

The great majority of fossils are formed in a water environment. This is because creatures can be buried rapidly in the muck and ooze at the bottom of a lake or stream, and become fossilized in that oxygen-free environment. Small size is also a favorable factor in fossil formation, since a small creature can be entombed more easily than a large one. A hard, bony organism is also favored over a soft, fleshy one. Fossils of shells are much more common than those of butterflies, moths, or jellyfish. Thus, the fossil record is imperfect in many ways. It favors many creatures over others. Yet it provides the best record we have of prehistoric life.

HOW THE AGES OF ROCKS ARE DETERMINED

Determining the age of a fossil or a rock is much more complicated than it sounds. But it is a fascinating subject that we will spend a few minutes examining.

There are two general methods of establishing the ages of rocks or other material. The first is called *relative dating*. Techniques of relative dating can provide an ordering or sequencing of the rocks, but cannot provide absolute ages of things.

Absolute dating, on the other hand, provides techniques whereby the exact age of an object can be determined. You might ask why relative dating is even necessary, if techniques of absolute dating can tell you exact ages. The answer is that absolute dating is not possible for all materials, and where it is possible, it may not be as reliable as relative dating. In practice, the two techniques are used together. A general sequence is first determined through relative dating; next, appropriate samples are selected on which to apply techniques of absolute dating.

Relative Dating

Almost all relative dating is performed on sedimentary rocks. These rocks are deposited in layers, with the oldest layers on the bottom and the newest on the top. This extremely simple "bottom-to-top" idea is complicated by tectonic activity that can contort the boundaries between layers. The principle of bottom-to-top still applies, but some work may be necessary to identify the true bottom and top.

The Grand Canyon is a magnificent example of rock that can be analyzed by relative dating. The action of the Colorado River has exposed the layers of sedimentary rock deposited over long periods of time.

A technique called *correlation* is used to match layers of rocks in different areas. For example, if fossils of the same animals are found in two separate locations, we can sometimes infer that the associated layers of rock came from the same points in time. But this is tricky. Animals can migrate and change their habitats over time, so care must be taken.

The final goal of correlation is to obtain a result called the *geologic column*. The geologic column is an abstract concept. It's what would result if we could take all the world's rock layers, order them with respect to each other, and then arrange them in a tall column with the oldest layers on the bottom and the newest on the top.

Even though the geologic column is an abstraction, we imagine it exists and divide it into regions that contain similar fossil forms. Lines are drawn between regions where significant changes seem to have occurred. Major divisions are called *eras*, and minor divisions within the eras are called *periods*. Figure 4-1 is a rough sketch of the geologic column.

The demise of the dinosaurs marks the end of the Mesozoic Era and the start of the Cenozoic Era, in which we now live. The last period of the Mesozoic is the Cretaceous, and the first period of the Cenozoic is the Tertiary. The boundary between the Cretaceous Period and the Tertiary Period is referred to as the "K-T boundary." The K-T boundary, which can be located in rock layers throughout the world, marks the point in time when the dinosaurs died off.

The present Cenozoic Era is divided into two periods, the Tertiary and the Quartenary. But smaller divisions, called *epochs*, are used when referring to the present Cenozoic. Thus, we often speak of the Cenozoic as divided into the Paleocene, Eocene, Oligocene, Miocene,

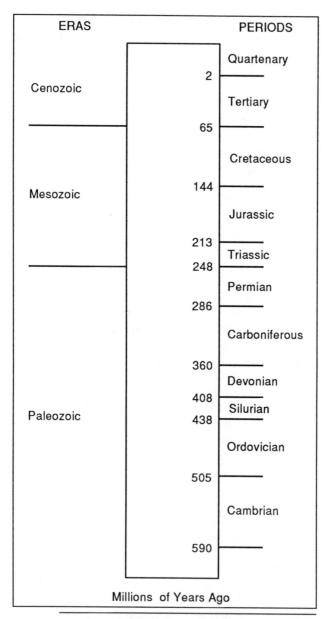

FIGURE 4-1. *The Geologic Column.*

Pliocene, Pleistocene, and Holocene Epochs. The Pleistocene Epoch represents the time of the most recent ice age. The Holocene represents the time from the end of the last ice age to the present.

Absolute Dating

Objects that contain some amount of radioactive material can often be dated absolutely. The method depends on the long-term statistical reliability of radioactivity. An analysis of a sample containing radioactive elements can often be used to determine the length of time that the radioactivity has been occurring. This is equivalent to determining the age of the sample.

Dating by this method obviously requires that the sample contain radioactive material. The speed at which radioactive decay occurs for those materials must also be well understood. Thus, this method of dating cannot be applied to every sample.

Radioactive decay is the process in which one chemical element systematically changes into another. A certain chemical isotope may contain too many neutrons, and thus is not able to hold itself together. Neutrons, protons, and electrons are released by the unstable element. In the process, the unstable chemical element changes, over time, into another element. (This process has already been described briefly in Chapter 3.)

The process of radioactive decay is gradual. All atoms in a sample do not release their protons, electrons, and neutrons simultaneously. A certain portion does over periods of time.

The process can be compared to human beings dying on Earth. The number of people on Earth is extremely large. Each day, a certain proportion die. It is certainly not the same number each day. But, in approximate terms, the number of people who die on any day is a percentage of the total number of those still alive.

Radioactive decay is the same idea, but mathematically a little simpler. This is because radioactive atoms in a sample do not get born. They only die. The number that die in a certain interval of time is proportional to the total living population at that time. Over years, the amount of the radioactive material (called the "parent") decreases, and the amount of new material (called the "daughter") increases. The decrease of the parent and corresponding increase of the daughter follow predictable mathematical patterns, depending on the chemical elements involved.

The speed at which radioactive decay occurs in an element is measured in terms of the *half-life* of the element. The half-life of a radioactive element is the amount of time required for the amount of radioactive material to be cut in half.

Absolute dating can be attempted on objects that contain radioactive materials of known half-lives. The idea is to measure the ratio of the parent atoms to the daughter atoms. If we know how many daughter atoms were created and how many parent atoms are left, we can use mathematics and the known half-life to calculate the number of years radioactive decay has been occurring.

But there is a catch. We need to assume that, when the whole thing started happening, there were no daughter atoms present. If daughter atoms were present at the start, then we need to estimate how many there were. Otherwise, we won't get an accurate number of how many were created by radioactivity.

Estimating the amount of daughter atoms initially present requires that we analyze several similar samples of different dates. We can look, for example, at similar rocks of different ages and assume that their chemical composition is very similar. We can use the mathematical concept of "simultaneous equations" to estimate simultaneously the ages of each of the rocks and the initial proportion of the daughter element, which we assume to be the same for all the rocks. The process is greatly simplified if we can estimate the age of one of the rocks using techniques of relative dating.

Carbon-14 is a common element used for absolute dating. It is particularly useful because it exists in all living material in essentially the same initial proportions. Thus, it provides an especially simple means for absolute dating, since daughter elements do not need to be counted. The age can be determined directly from a measurement of the proportion of carbon-14 in the sample.

Unfortunately, as always, there are a few catches. The first has to do with the half-life of carbon-14, which is about 5,700 years. With this comparatively short half-life, it is difficult to date objects more than about 60,000 years old, because the proportions of carbon-14 in these objects will be extremely low and not measurable.

The second catch is that carbon-14 is what is called a *cosmogenic* isotope. This means that it is produced by cosmic rays from the sun, which bombard other carbon atoms with neutrons. As solar sunspot activity increases, more cosmic rays are deflected by the sun's magnetic field, and less reach the Earth to help create carbon-14 atoms.

The second catch is actually a plus, but it is a tricky plus. The amount of carbon-14 produced during any one time depends on the amount of solar activity. At times of solar minima (when fewer sunspots occur), more carbon-14 is created and living organisms contain slightly higher proportions. The catch, therefore, is that carbon-14 dating is only completely accurate if we know the history of solar maxima and minima. The plus is that the measurement of carbon-14 in samples can yield information about the sun's history.

Another important cosmogenic isotope is beryllium-10, which decays with a much longer half-life of about 1.5 million years. Beryllium-10 is present in ancient ice that can be extracted in cores from

Antarctica or Greenland. Its presence in the ice cores can be used to determine the history of solar fluctuations, as can the presence of carbon-14 in organic material.

A particularly important radioactive element is potassium-40, which decays into argon with a half-life of 1.25 billion years. The exceptionally long half-life makes it possible to date anything during the entire history of the Earth that contains radioactive potassium-40. An additional advantage is that the daughter element argon almost never occurs naturally in materials that are being dated, so the presence of argon in these materials is usually due solely to radioactive decay.

There are a number of other radioactive isotopes used for dating. Uranium-238 decays into lead with a half-life of about four billion years, so it too provides a useful long-term dating method.

When methods of relative dating and absolute dating are used together, they are effective checks against each other. Current state-of-the-art technology can measure amounts of parent and daughter isotopes to a fraction of a gram. The complex process of determining the age of a rock can be done quite accurately with the help of modern tools.

THE FACTS OF THE CASE

"What killed the dinosaurs?" is such an alluring mystery that almost everyone seems to have an opinion. Many of the proposed theories are questionable in light of the full set of known facts.

Thus, the theory that the dinosaurs were destroyed by a new breed of egg-eating mammals can immediately be challenged, since it does not explain the simultaneous extinction of marine dinosaurs.

Another theory is that the dinosaurs were poisoned by flowering plants, which contained toxic materials. This theory can be challenged, since flowering plants appeared about forty million years before the end of the Cretaceous Period, when the dinosaurs died. A third theory posits that a giant explosion of a star, called a *supernova*, occurred, showering Earth with lethal doses of radiation. The theory can be challenged, because a mathematical calculation shows that the star would have to be less than one light year away to account for the observed damage. Presently, the closest known star to us is about four light years away. The chances of a star at some point in time being less than one light year from Earth and happening to explode in a supernova are essentially zero.

Another theory is that a reversal of the Earth's magnetic field occurred at the time. This happened somewhat gradually, so the Earth's magnetic field weakened immediately before it reversed. With the weaker magnetic field, cosmic rays could not be deflected by Earth, and the doses became lethal. The theory can be challenged by simply examining the record of magnetic reversals, and finding that none coincided with the death of the dinosaurs. Moreover, no known mass extinctions have coincided with known magnetic reversals.

Clearly, we must learn the facts of the case before we can propose a theory or evaluate the existing theories intelligently.

An interesting approach is to look at the problem as the police do when they attempt to solve a murder case. Has more than one murder been committed? Do similarities in the *modus operandi* (way of operating) point to the same murderer? What is the nature of a murderer who would use such a *modus operandi*?

Historical Patterns of Extinctions

The great extinction of the dinosaurs sixty-five million years ago is only one of many mass extinctions that occurred during the Earth's history. Generally, extinctions accompany the imaginary dividing lines in the geologic column. These dividing lines were drawn to separate periods in which the fossil record changed significantly. Such changes normally imply the extinction of several species and the replacement by others.

By examining the fossil record and noticing these kinds of changes, it is obvious that mass extinctions have occurred many times throughout history. The extinction of the dinosaurs at the K-T boundary was one of a long series of deeds, possibly all by the same serial murderer.

What did these extinctions have in common? The answer is quite a lot. In all cases, animals on the land and in the sea were affected. But freshwater fish were not affected at all, only oceanic fish and land animals.

Small animals were much less affected than large animals. Plants were killed, but mostly in the temperate and subarctic zones. Tropical plants were not affected.

In the K-T boundary extinction, crocodiles and sharks survived. Birds survived, but not the great pterosaurs. The puzzle deepens. What sort of killer can this be, that strikes repeatedly with these kinds of patterns?

When did the killer strike? In terms of millions of years ago, major extinctions took place at the following times: 249, 211, 191, 175, 160, 137, 125, 91, 65, 36, 14. Is there a pattern to these numbers? Some scientists think so. Most of the times between extinctions lie somewhere between twenty and forty million years.

Possible errors in the data are significant. Each one of the ages in the series has a likely error of up to five million years in either direction. Thus, the extinction that took place about 137 million years ago actually may have taken place sometime between 132 and 142 million years ago.

By throwing out extinctions that are not as large as the others and making constructive use of the possible errors, some believe that the extinctions follow a periodic pattern. According to this idea, every twenty-six to thirty million years, a major extinction occurs.

Not all scientists agree with this view. Some think that the periodicity of extinctions is the result of imaginative data manipulation and nothing more. Others point to advanced statistical methods indicating that such a pattern, if not periodic, would only occur by chance with a probability of 10%. The debate is intriguing. If, in fact, the extinctions are periodic, what could be the nature of such a killer that would strike with this kind of regularity?

Theory formation is a creative process. One who formulates a theory must specify an event or event sequence that is both consistent with all known facts and sufficient to explain the facts. At the same time, the theory must be complete; it must not leave a trail of unexplained loose ends. The theory should also provide the most likely explanation that meets the first three conditions.

In the case of the dinosaur killer, it is debatable that there is any one theory currently existing that is consistent with all the facts. There are three theories we will look at in this chapter. In each case, it is debatable whether or not the theory is consistent with all the facts. But this much is true: None of them can be refuted outright.

Of the three theories we will discuss, two belong to the school of "catastrophism," and one belongs to the school of "gradualism." The impact theory and the volcanic eruption theory are the two catastrophes. The continental drift theory is the representative for gradualism.

THE IMPACT THEORY

According to this theory, Earth was struck by a huge object from outer space—a comet, an asteroid, or a meteorite. The resulting explosion filled the atmosphere with so much dust and vaporized rock that the sun was effectively blotted out for several months. Photosynthesis in the plants could not occur with insufficient sunlight. Many of the plants died as a result. Following the death of the plants came the death of the plant-eaters, who fed on the plants, then the death of the carnivores, who fed on the plant-eaters.

Crocodiles and sharks, who were able to live on the meat from corpses, were among the few to survive. Mammals, smaller and more resilient than the dinosaurs, were able to stick it out through the hard times. They required less food and were less sensitive to sudden temperature changes than the dinosaurs.

We will briefly consider the three types of objects from outer space that could have hit the Earth.

Comets

Most scientists believe that comets are as old as the solar system, about 4.6 billion years. They are the most pristine objects in our part of the galaxy, consisting of the same materials of which they were formed.

The sun and planets of the solar system probably formed from a large, spinning, disk-shaped cloud of gas called the solar nebula. Over time, the gas collapsed through its own gravity. The sun formed near the center. Planets formed in areas away from the center, in a process called *coagulation*. Minute dust grains struck each other in random collisions, sticking together because of the combination of van der Waals forces and gravity.

Eventually, they become big enough to form the planets. Planets all orbit the sun in the same plane, due to the original shape of the solar nebula, which resembled a flattened pancake.

It is believed the comets are leftover debris from the process of planetary formation. They are collections of coagulated dust that never became big enough to form planets.

Comets most likely originated in a region called the *Kuiper Belt*, an area roughly bordered by the orbits of Uranus and Neptune. Sizes of comets vary from less than one mile in radius to perhaps as big as sixty miles or more. Halley's comet was about 5 x 5 x 10 miles along its dimensions. Most comets probably have radii in the range of one to five miles.

As such, comets are very light objects when compared to stars and planets. They are thus strongly affected by the gravitational pull of these objects.

When a comet passes in front of a planet, the gravity of the planet acts to slow down the comet. The comet's orbit about the sun shrinks as a result. On the other hand, when a comet passes behind a planet, the gravity of the planet acts to speed up the comet. As a result of this energy gain, the comet's orbit grows larger. Thus, in the process of moving about and minding its own business, the orbit of a comet can change significantly, depending on its approaches to large planets.

These interactions not only affect the size of the comet's orbit but also its shape. Depending on which direction the comet is moving when it encounters the planet, its orbit may become more elliptical or more circular. The mathematics of this effect are complex. A simplified approximate statement would be to say that comets that gain energy while they are moving away from the sun, or that lose energy while they are moving toward the sun, will see their orbits stretched. Comets that lose energy moving away from the sun or that gain energy moving toward the sun will see their orbits become more circular.

Another effect on the orbits of comets comes from the force of the central plane of the Milky Way Galaxy. This huge mass lies "below" (southward of) the solar system, and is inclined to the plane of the solar system at an angle of about sixty degrees. The central plane affects comets the most when they are farthest from the sun. At these points of the comets' orbits, the force of the sun is not so great, and they are more easily pulled in other directions by the galaxy's central plane. The small masses of comets also make them more vulnerable to such perturbations from entities outside the solar system. Forces from the central plane throw the orbits of comets askew (out of the same plane) to the rest of the planets in the solar system.

Over the course of billions of years, comets' orbits have been changed dramatically by interactions with planets, forces from the central plane of the galaxy, and forces from stars that pass by at random. As a result, most comets have long since been expelled from their birthplace, the Kuiper Belt.

The term *Oort Cloud* is now used to define the region of space where comets spend most of their time. It is a huge egg-shaped region surrounding the sun and extending to a maximum distance of about two light years. The Oort Cloud is a gigantic oval, with its long axis inclined about sixty degrees to the disc-like plane of the sun and its planets. Statistical methods have been used to estimate the total population of comets in the Oort Cloud. Estimates range from one trillion to about 1,000 trillion.

Halley's comet, which orbits the sun with a period of about seventy-six years, is a rare example of a comet that still resides in the Kuiper Belt. Its orbit has been stretched considerably, but it never wanders outside the realm of the solar system and its planets.

Comets were formed in the outer section of the nebula out of which the solar system grew. At this time, most of the denser, heavier elements gravitated toward the middle of the nebula. Therefore, most of the chemical elements found in comets are lighter ones. Typical elements making up comets are hydrogen, oxygen, nitrogen, carbon, and sodium.

Traces of heavier elements such as silicon, sulfur, potassium, and calcium are also there, but in less abundant quantities. A large portion of a comet's mass is water ice, and it has been theorized that cometary impacts with the Earth led to the formation of the Earth's oceans.

The tails on comets are caused as the comet approaches the sun and heat from the sun causes the water ice to *sublime*, that is, change directly from ice to gas, a phenomenon that is possible at the extremely low pressures of outer space. The expanding gases drag dust particles away from the comet.

The dust particles are then affected by two forces from the sun. First, is a gravitational force, which pulls particles toward the sun. The force on any one particle is proportional to its mass. A second force is due to radiation pressure, and tends to push dust particles away from the sun. The force on a particle due to radiation pressure is approximately proportional to the particle's area. Thus, some particles of a comet are pulled toward the sun to form a "head," and some are pushed away from the sun to form a "tail."

Asteroids

Asteroids are chunks of rock that lie mostly between the orbits of Mars and Jupiter. They range in size from the largest (Ceres), which is about 600 miles in diameter, to tiny rocks. So far, about 4,000 asteroids have been catalogued, but many more certainly exist. Most asteroids that we have found are at least twenty miles in diameter.

Asteroids are remains of the early solar system. They are rocks that were unsuccessful in sticking together to form a planet. Probably, strong gravitational forces from nearby Jupiter kept tugging on the rocks, pulling them apart and retarding accretion.

Asteroids were formed in the inner solar system, and hence are composed of heavier elements than the comets. They are stony objects, consisting of metals such as nickel, iron, and others.

Asteroids are deflected from their natural orbits around the sun through the gravitational interaction with Jupiter and a phenomenon known as *resonance*. An asteroid will revolve about the sun in a shorter amount of time than will Jupiter. Once in a while, it will pass Jupiter. As it passes Jupiter, the planet will tug on it, changing its orbit slightly.

How will these tugs affect the asteroid over time? If the tugs occur at different places each time, they will cancel each other out. If the tugs occur at the same place over time, the tugs will "add up," and the orbit of the asteroid will become distorted. It will move in more of an elliptical pattern, which in turn allows it to be affected by the inner planets as well. Over time, its orbit is affected so much by other planets that it bears no relation at all to its original motion between Mars and Jupiter.

The resonance phenomenon, which here means repeated tugs by Jupiter at identical locations, has been verified observationally. If the period of rotation about the sun of each known asteroid is listed, interesting missing pieces called *Kirkwood gaps* occur. None of the asteroids with certain periods of rotation are observed. These periods of rotation in which the gaps occur are directly related to the period of rotation of the planet Jupiter. Asteroids whose periods are certain ratios of Jupiter's period are not observed. Some of these critical ratios that define the Kirkwood gaps are 2:1, 3:1, 4:1, and 5:2.

Asteroids that at one time lay in the Kirkwood gaps, now are flying about the solar system in essentially random, haphazard orbits. Such asteroids threaten to collide with Earth.

Meteorites

The term *meteorite* refers to a rock from space that enters the Earth's atmosphere, survives the trip, and crashes to the ground. The term *meteoroid* refers to any rock in space. There is a huge overlap, then, between meteoroids and asteroids, at least according to the definition.

In fact, about 99% of meteoroids and meteorites are really asteroids, or asteroids that have broken up. Most of the rest are of lunar origin, a fact that can be shown by chemical analysis. This class is probably formed by collisions of large meteorites with the moon, in the process shooting lunar debris upward at a high enough velocity to escape the moon's gravity.

Another interesting class of meteorites appears to have come from Mars. Potassium-argon dating of these meteorites shows that they are very young, much younger than the age of the solar system. Therefore, they cannot be asteroidal in nature. Evidence of volcanic origins rules out the moon. Debris from Mercury or Venus would most likely orbit the sun in a much tighter orbit, and never make it out as far as the Earth. Mars seems the only possible origin. Scientists believe that the meteorites from Mars were probably caused by the impact of an asteroid with Mars. The only other possibility, a volcanic eruption, would probably not be able to hurtle rock outward with a velocity of three miles per second, the speed necessary to escape Mars's gravity.

Evidence for the Impact Theory

Geologists examining rocks in the Earth have noticed an anomaly at the K-T boundary, marking the end of the Cretaceous Period and coincident with the death of the dinosaurs. The anomaly is an unusually high level of the heavy element known as iridium. The iridium at the K-T boundary has been found in sedimentary rock layers all over the world.

Iridium is very rare on the Earth's surface. This is because it is quite dense. As the Earth formed, almost all the iridium sank toward the Earth's interior. But iridium is a common element in extraterrestrial objects such as asteroids. Thus, the impact theory was born. The high iridium content observed throughout the world at the K-T boundary is possibly due to an impact of a heavenly body with the Earth.

Chemical analysis of the boundary layer clay not only shows iridium present, but existing in the same proportion to other elements as is displayed in meteorites. So it is not just the presence of iridium, it is its presence in exactly the right proportion.

More evidence comes from the presence of mineral spherules, also found in the clay at the K-T boundary. The spherules are small pieces of basaltic rock that were shock-melted upon impact with the Earth. The basaltic nature of the spherules suggests that the impact may have occurred in an ocean, since oceanic crustal rock is basaltic in nature. Spherules have been noticed as resulting from known meteorite impacts on the land, but their chemistry is not basaltic.

Shocked grains of quartz, possessing bands of deformation, have also been dug out of the clay at the K-T boundary. Such grains are very rare, and have only been seen at impact craters and nuclear test sites.

A possible "smoking gun" has also been identified. Some scientists believe that an impact occurred about sixty-five million years ago in the Caribbean Sea, near Mexico's Yucatan Peninsula. The iridium anomaly at the K-T boundary in nearby Haiti is particularly pronounced, confirming the possibility that an impact occurred there at about the time that the dinosaurs died.

The Nemesis Idea

Some scientists have tried to link the observed pattern of periodic extinctions with the impact theory. The result is an imaginative and fascinating idea known as the "Nemesis" or "death star" theory.

A list of all known impact craters and their ages has been assembled. It is quite difficult to date impact craters, and the possible errors are large. Some argue that impacts, as mass extinctions, occur in cycles. This claim is extremely controversial, however.

It all depends on statistics, which can be a tremendously tricky subject. How do you account for possible errors in the dating of craters? How do you account for the fact that most datable impact craters are biased toward the more recent times?

But if we disallow these objections for the time being, one set of data suggests that impacts occur in peaks with a cycle time between peaks of about thirty-two million years. It is tantalizingly close to the extinction periodicity cycle, which lies somewhere between twenty-six and thirty million years. Moreover, the series of peaks of cratering roughly corresponds to the dates of known extinctions.

Could all this just be chance? Or is there really some hidden mechanism that causes impacts to occur in cycles?

One possibility involves the central plane of the Milky Way Galaxy. The sun bobs up and down through the central plane, describing simple harmonic motion. The period of the sun's oscillation is about sixty-six million years. Equivalently, the sun makes a passage through the galactic plane about once every thirty-three million years, or twice for each complete oscillation. This is very close to what we are looking for.

The idea is that, as the solar system passes through the central plane, the gravitational force of the central plane becomes very large and perturbs the orbits of many comets, throwing them into the inner solar system and endangering Earth.

Unfortunately, the sun's positioning relative to the galactic plane is out of sync with the extinction pattern. We are presently very close to the galactic plane. Thus, the galactic plane theory suggests we should be very close to an impacting peak, and that the last such peak occurred about thirty-three million years ago. But the last extinction was about fourteen million years ago. So this idea appears to be off the mark.

The next idea is the Nemesis theory. According to the Nemesis theory, the sun is not a lone star, but has a twin, a binary companion. The assumption is reasonable. Astronomers are discovering that most stars similar to the sun are actually binary stars, not single stars.

In a binary star system, the two stars rotate about their common center of mass. If the system has planets, the planets may rotate about the heavier star (if one is dominant) or describe very complex orbitals, such as figure-eights, about both stars if they are about the same size. The Nemesis theory suggests that the sun has a very dim binary companion, which is much smaller than the sun. In such a case, the sun's motion and the orbits of the planets would be little affected by the companion star.

The companion star's importance comes about when it passes through the Oort Cloud, as it surely must when it revolves around the sun. On passing through the Oort Cloud, the orbits of thousands or millions of comets would be perturbed in a somewhat random fashion. Some, by chance, would be hurtled into new orbits that would threaten the planets of the inner solar system, including Earth.

If the companion star orbited the sun every twenty-six to thirty million years, it could explain the observed pattern of extinctions. The theory is brilliant, creative, and among the most fascinating to come out of modern science today.

Objections to the Impact Theory

Much of the debate surrounding the impact theory turns on the accuracy of the fossil evidence. Fossil evidence suggests that dinosaurs died out gradually over a period of several thousand years. Such a prolonged dying off is obviously inconsistent with the impact theory, which would call for a rapid mass extermination.

Proponents of the impact theory point to inadequacies of the fossil evidence, as discussed earlier in this chapter. Proponents also point toward the fossil record of certain marine organisms known as foraminifera. The fossil record of these creatures shows that they died off very abruptly, at the end of the Cretaceous Period. Furthermore, the fossil record of these creatures should be accurate, since they are, upon death, quickly covered by sediment and entombed in an anaerobic environment.

Opponents counter, saying that it is a major achievement to date a fossil to within 10,000 years. It is thus impossible to say how quickly the foraminifera died off. The evidence is also consistent with the hypothesis that they disappeared gradually. Ten thousand years, although a very brief amount of time in the geologic sense, represents a significant fraction of the time that modern civilization has existed, to use a comparison. It is just not possible to conclude with reasonable certainty that the marine organisms died off quickly, given the inherent errors in dating rocks.

Another objection has to do with the plant life that died off. Since tropical plants are more in need of sunlight than their higher latitude counterparts, we would expect tropical plants to be more affected by a blotting out of the sun's radiation. Yet the opposite occurred. Plants died en masse in North America, but escaped relatively unscathed in the tropics.

This objection to the impact theory seems to have no adequate response. This is not to say that the impact theory must be scrapped because of this single objection. But it has an unexplained loose end.

One objection to the Nemesis theory has to do with the composition of comets. The element iridium is a heavy metal found in asteroids and meteorites. Its presence in comets has not been verified. If it is present in comets, it would be only in minute quantities. It is very doubtful, therefore, that the impact of a comet or even a series of cometary impacts could have led to the pronounced iridium anomaly at the K-T boundary.

This does not mean that the Nemesis theory is dead. Cometary impacts could still occur in peaks, and occur coincidently with the major extinctions. But there appears to be no way to link a cometary impact with the iridium anomaly.

A final argument made by impact theory opponents is that there is no evidence to show that an impact actually did kill the dinosaurs. The iridium anomaly may be strong evidence that an impact occurred, but it does not prove that the impact killed the dinosaurs.

Supporters counter by saying that it is too much of a coincidence that an impact would occur at the exact point in time when the dinosaurs disappeared forever. Opponents counter, saying that dinosaurs were on the decline long before the alleged impact. Coincidence does not prove causality.

VOLCANIC ERUPTION THEORY

An alternate catastrophe theory says that we don't need extraterrestrial sources to explain the iridium anomaly, the shocked quartz, or the mineral spherules. All these clues are consistent with extensive volcanic activity.

Studies have shown high amounts of iridium emitted by the Kilauea volcano in Hawaii, and also by the Piton de la Fournaise volcano on the island of Reunion in the southern Indian Ocean near Madagascar. In addition, viscous magma leads to explosive eruptions and shock features consistent with the shocked quartz and mineral spherules found in the clay of the K-T boundary. Evidence of such shock has been discovered in rocks from some more recent violent volcanic eruptions. One example is the eruption of the large Toba, Sumatra volcano of Indonesia, which erupted violently about 75,000 years ago.

Massive and prolonged volcanic activity would affect the climate worldwide in a traumatic fashion, as we saw in Chapter 3. This is the heart of the Volcanic Eruption theory. Prolonged eruptions affected the climate traumatically, killing many species of plants and animals before they were able to adapt to the quick changes.

But what would induce such periods of massive volcanic activity? The theory offers an ingenious explanation, linking the two apparently unrelated ideas of the Earth's magnetic field reversals and hot spot volcanoes.

The Earth normally reverses its magnetic field about once every 250,000 to 1,000,000 years, but there is considerable variation in this interval. A noteworthy deviation occurred during the Cretaceous Period, when the magnetic field experienced no reversals for thirty-five million years. The period extended from about 120 million years ago to about eighty-five million years ago. Another long period when the Earth experienced no magnetic reversals ended about 250 million years ago, shortly before the largest mass extinction known. Coincidence?

It is not known for sure what forces in the Earth cause the magnetic field to reverse. But the process is certainly related to activities that build the magnetic field in the first place. These activities are the movements of liquid iron through the Earth's interior. If the liquid iron deep in the Earth is moving freely, magnetic reversals will occur periodically. In this sense, magnetic reversals are a measure of the amount of activity in the region known as the Earth's outer core, where the liquid iron resides.

If the liquid iron is not moving freely but is restricted in some way, magnetic reversals will not occur. In such a state, heavy buildups can occur in certain areas because of blockages inhibiting the free movement of the liquid iron. The heavy buildups of liquid iron in concentrated clumps transfer heat upward unevenly. A volcanic hot spot is formed above such a region. Continuous high heat from the concentration of liquid iron radiates upward through the Earth's mantle. Large amounts of magma are formed, which drift upward, forcing their way through the Earth's crust and forming hot spot volcanoes. This is the theory.

Evidence for the Volcanic Eruption Theory

A hot spot existed at the time of the death of the dinosaurs. It is known as the Deccan Traps and is in present-day western India. Due to plate movements, the hot spot corresponding to the Deccan Traps lies at present under the island of Reunion in the southern Indian Ocean. An interesting but probably meaningless observation is that the Reunion Island hot spot is almost exactly opposite the Hawaiian Islands hot spot on the surface of the Earth. A straight line passing through the center of the Earth and going through Reunion would come out at the other end very close to Hawaii.

Besides the Deccan Traps, which correspond to the mass extinction of sixty-five million years ago, other volcanic hot spots have been identified that correspond to other mass extinctions. Violent hot spots have been identified that correspond to extinctions at the following times (in millions of years ago): 249, 211, 191, 137, 91, 65, 36, and 14. In most, but not all, cases, the formation of the hot spot was preceded by a long period of no magnetic reversals.

There is some evidence showing that the layer of iridium-rich clay at the K-T boundary was spread over a period of about 500,000 years. This spread is consistent with a prolonged period of massive volcanism, but is not consistent with a single asteroidal impact. The shocked quartz grains and mineral spherules at some geological sites exhibit the same distribution spread.

Analysis of the Deccan Traps indicates that they were created during a relatively short period of time of intense volcanism. The date of the volcanic activity has been set at between sixty-four and sixty-eight million years ago, corresponding closely to the K-T boundary extinction. Some magnetic and fossil studies lead to the conclusion that the Deccan Traps were created within a time interval of 500,000 years or less.

Some scientists have pointed out that extinctions at the end of the Cretaceous Period were not instantaneous, but exhibit a type of stepwise pattern of extinction and recovery. The pattern is consistent with massive and prolonged volcanic activity.

The chemical composition of the Earth's mantle, it turns out, is very similar to the chemical composition of meteorites. Thus, the chemical composition of the clay layers at the K-T boundary can be explained either by an asteroidal impact or by volcanic activity. The case for the volcanic eruption theory appears strong.

Objections to the Volcanic Eruption Theory

Much of the foundation of the volcanic eruption theory turns on the interpretation of the fossil record. Most scientists agree that a single volcanic eruption could not kill the dinosaurs. Volcanic activity would have to persist over many thousands of years to create the necessary effects. The inherent inaccuracies in dating fossils rule out a definitive statement, at least at this time and with existing technology. We simply can't say for sure whether the dinosaurs died out very quickly or very gradually. For the volcanic eruption theory to be acceptable, the extinctions would have to take place over just the right interval of time—not too short, because the volcanoes need time to work, and not too long, because massive volcanic activity cannot be maintained for longer than certain lengths of time, given the physical limits of the Earth's internal machinery. Thus the volcanic eruption theory cannot at the present time be validated.

The volcanic eruption theory is incomplete. The mechanism describing the creation of volcanic hot spots has not been verified. Not every major volcanic hot spot has had its origin after such a time period absent in magnetic reversals. So the connection between an absence of magnetic reversals and the creation of a volcanic hot spot is imperfect, at best.

Many of the objections to the impact theory also apply to the volcanic eruption theory. The fact that tropical plants survived better than temperate or subarctic plants is inconsistent with either theory.

Even if we can prove with certainty that massive volcanic activity coincided with the dinosaur extinction, we still cannot prove that one caused the other. Maybe massive volcanism took place and maybe it didn't. If it did, then maybe it killed the dinosaurs, and maybe it didn't. There are a lot of "maybes" and a lot of "ifs."

THE CONTINENTAL DRIFT THEORY

During the time that the dinosaurs flourished, there was only one landmass on the Earth. All the continents we know today were connected together into a massive supercontinent known as *Pangaea*. Large portions of Pangaea existed in tropical areas. This circumstance proved to be a great two-edged sword.

On the one hand, the dinosaurs were free to migrate wherever they wished. If they didn't like the climate or availability of food where they lived, they simply moved.

Of course, dinosaurs did not have the intelligence to be able to figure out where to go to find a warmer climate or more food. Their migration was random. Some dinosaurs, by chance, moved to favorable regions and prospered. Others moved to unfavorable regions and died. But for many millions of years, the possibility of migration existed. At the same time, the seeds for the dinosaurs' extinction were laid by this continental configuration. Competition among the dinosaurs existed over the most favorable regions of the planet. Dinosaurs roamed from afar to find the favorable spots. When they reached the favorable spots, either they had to fight to stay there, or they had to move to the unfavorable spots.

As the Cretaceous period moved onward in time, the richness of the dinosaur life declined. A few species began to dominate. The phenomenon can be mapped well by examining the fossil record and using a type of measurement known as the *Simpson index*.

The Simpson index is a measure of the richness of an ecosystem. It is based on the following thought experiment: If you select two members of the population at random, what is the probability that they will be relatives? (Biologists define my rather loose term "relative" as belonging to the same genus. An organism's genus is one step up from its species in the scheme of biological classification.)

If we calculate the value of the Simpson index for periods of time throughout the Cretaceous period, we notice a definite decline in the richness of life as the period progresses. Clearly, something was happening. Dinosaurs were in trouble ecologically long before the time of their final disappearance at the end of the Cretaceous.

The lack of richness and variability of dinosaur life set them up for the final blow. As dinosaur life developed when the continents were stuck together, certain species were favored over others by very poor and narrow criteria, none of which was the ability to withstand climate change.

If more species of dinosaurs had been around at the end of the Cretaceous, many more of them would have survived. Many would have randomly evolved so they could withstand the changes that took place at the end of the Cretaceous. But with a small number of dominant species around, the enormity of the disaster was magnified. With the exception of crocodiles and birds, the dinosaurs were wiped out because of their inability to adapt to change.

The forces of plate tectonics forced the continents apart and generally northward toward cooler weather. Migration routes were cut off as land masses separated. Dinosaurs became stranded on whatever piece of land happened to be carrying them. As their continents drifted northward, they had no choice but to succumb to the cold.

Mammals were much more fortunate. Most of this was nothing more than lucky timing. Mammals came into being as continents were drifting apart. They spread across the world rapidly. Their primary stages of evolution occurred while they were spread out, occupying regions of vastly different climates. The mammals evolved into a very rich population, as measured by the Simpson index. When they were forced to adapt, many of them were able to make the adjustment. They experienced climate change early in their evolutionary history, and this made the difference.

When the big climate change hit, much of it was due to the northward drift of continents toward cooler regions. Another part of it came from the dawning of the first ice era in over 150 million years.

Ice eras are defined as periods in the Earth's history when glaciation exists. We are currently living in an ice era, the same one that began sixty-five million years ago.

Ice eras are caused by particular continental configurations. When circulation of ocean water between the tropics and the polar regions is blocked, glaciation can occur in the polar regions. Today, central Antarctica is surrounded by land, and thus cannot be warmed by circulating ocean water; similarly, the North Pole is almost completely surrounded by land, so circulation of ocean water to it is severely restricted.

About sixty-five million years ago, the continents reached a configuration that approximates what we see today. An ice era started, where glaciers began to form in the polar regions.

The ice in glaciers has a very high *albedo*; that is, it reflects most of the solar radiation that reaches it. Thus, the very presence of ice tends to make the Earth colder still, since the Earth as a whole absorbs less of the heat that reaches it.

Another effect of glaciation is the lowering of sea levels and the draining of shallow seas at the edges of continents. As glaciers form, they take their share of the water out of the world's ecosystem. The effect of this is seen in falling ocean levels.

Falling ocean levels result in the disappearance of warm ocean water. Shallow seas, which are significantly warmer than the neighboring deep ocean water, drain off and mix with the colder oceans. Marine plants and animals that had adapted to the warmth of the shallow seas must make a brutal adjustment. Of course, the oceans on the whole become colder due to the glaciation-induced climate change. All creatures living in the oceans become vulnerable, particularly plants and other animals that are unable to migrate.

Organisms inhabiting freshwater lakes fare better than their counterparts in the oceans. Temperatures in freshwater lakes vary much more than temperatures in the ocean do through the natural course of a year. Therefore, organisms used to living in freshwater lakes were already somewhat prepared for temperature fluctuations.

The continental drift theory, it can be argued, is the only theory that is consistent with all the known facts. Unlike the volcanic eruption theory or the impact theory, the continental drift theory explains why temperate and subarctic plant life fared worse than tropical plants.

The climate changes resulting from continental drift and subsequent glaciation would be most strongly felt in the polar region, closest to the glaciation.

Opponents of the continental drift theory claim that it doesn't account for the observed periodic behavior of extinctions. Supporters of the theory have an array of arguments. Because several extinctions occur, perhaps in a cycle, does not imply that all of them were caused by the same thing. Of course, if you believe in the cyclical nature of extinctions, it is compelling to think that they come from a common cause. If you don't believe extinctions are cyclic, then anything could be causing them, including different things each time. Certainly, the inference that extinctions come in cycles is questionable, given the data.

Whichever theory you believe depends on how you interpret the fossil evidence. If the mass extinction occurred very suddenly, then the impact theory becomes the favorite. If the extinction occurred in a stepwise fashion over several thousand years or perhaps half a million years, then the volcanic eruption theory is the favorite. If the extinctions were gradual, then the continental drift theory seems to be the only acceptable answer of the three.

CONCLUSIONS

Well, here it comes—the one time in the book where I forget about scientific objectivity and just come out and say what I think. And so, this is it. Here is my own opinion of what it was that killed the dinosaurs.

You may have heard this interesting little puzzle. A man comes home from work. He gets to his apartment building, where he lives on the third floor. He steps into the elevator. He pushes the button that takes him to the second floor. He gets off and walks up one flight of stairs to the third floor. He has been doing this for years. There is nothing wrong with the elevator. Lots of people every day take it to go to the third floor or higher. Why does this person act like this?

Let's imagine that this is our mystery. The only facts we have are those presented in the previous paragraph. Let us now go into the creative process of theory formation. What circumstances that are consistent with all the facts can explain what's going on?

I maintain that there are two competing theories to explain this. I'm sure you, the reader, can think of more. But, for the sake of proving my point, let's keep it simple and say there are two theories.

The first theory goes like this. The man is a dwarf. The elevator buttons are arranged in ascending order, so that the button for the ground floor is on the bottom, above it is the button for the second floor, and so forth. The dwarf just happens to be of a height where he can barely reach the second floor button but not the third floor button. So he goes to the second floor. Let's call this the "dwarf" theory.

The second theory involves a man who feels he needs exercise. He works at a job where he sits most of the time. He believes he is not in very good shape, and he does not have the time to go to a health club or a gym to work out every day. He decides that every day he will walk up one flight of stairs to his apartment. He knows this is a small thing, but decides it is a good habit to get into. It is a simple thing to do that doesn't require much time. He starts doing it, and maintains the habit over several years. Let's call this the "exercise" theory.

Of these two theories, the dwarf theory is by far the more ingenious and creative. It captures our imagination because it is so clever. The exercise theory, by comparison, is dull and mundane.

But cleverness aside, which theory is the more likely to be correct? The probability that a given person is a dwarf is about one in 10,000. The probability that the elevator buttons exactly coincide with his height is probably about one in thirty. On top of that is the probability that the dwarf holds a conventional day job, the probability that he chooses to live in an apartment with this inconvenience, and the probability that he stays living there for several years. When you multiply together all these independent probabilities, you are looking at about one chance in a million that this theory is true.

The exercise theory is dull, but it is believable. We know there are a large group of men who have sedentary jobs, who know they need exercise and don't have time for it. The likelihood of this theory is much higher than that of the dwarf theory.

In the case of the dinosaurs, the continental drift theory is not very exciting. But it has one thing going for it. We know for sure that continental drift happened. We know by the alignment of rocks with the Earth's magnetic field through the ages. We don't have to theorize that maybe it happened.

The impact theory and the volcanic eruption theory are infinitely more clever. But they require theorizing. We don't know for sure if an asteroid hit the Earth, or if massive volcanic eruptions occurred precisely as the dinosaurs died.

We know for sure that prolonged climate changes, as were caused by continental drift, have caused extinctions. Many animals, such as the saber-toothed tiger, the great American ground sloth, and the woolly mammoth, who lived in the Ice Age, became extinct as the climate changed at the end of the Ice Age. On the other hand, we don't know for sure if an asteroid impact or volcanic eruptions can cause extinctions of the magnitude at the K-T boundary. The idea is reasonable, but we just haven't seen it happen. And so we really don't know. More theorizing is necessary.

As unexciting as it sounds, I believe that the man walked up the flight of stairs because he needed the exercise, just as I believe that drifting continents killed the dinosaurs.

BIBLIOGRAPHY

Books

Arduini, P., and Teruzzi, G. *Simon and Schuster's Guide to Fossils*. New York: Simon and Schuster, 1986.

Bakker, R. *The Dinosaur Heresies*. New York: Zebra Books, 1986.

Black, R. *The Elements of Paleontology*. 2nd ed. Cambridge: Cambridge University Press, 1988.

Chapman, C., and Morrison, D. *Cosmic Catastrophes*. New York: Plenum Press, 1989.

Chorlton, W. *Ice Ages*. Alexandria, VA: Time-Life Books, 1988.

Erickson, J. *Dying Planet: The Extinction of Species*. Blue Ridge Summit, PA: Tab Books, 1991.

Goldsmith, D. *Nemesis*. New York: Walker and Company, 1985.

MacFall, R., and Wollin, J. *Fossils for Amateurs*. 2nd ed. New York: Van Nostrand Reinhold, 1983.

Miller, R. *Continents in Collision*. Alexandria, VA: Time-Life Books, 1985.

Sagan, C. *Cosmos*. New York: Random House, 1980.

Sagan, C., and Druyan, A. *Comet*. New York: Random House, 1985.

Wilford, J. *The Riddle of the Dinosaur*. New York: Vintage Books, 1987.

Periodicals

Alvarez, W., Asaro, F., and Courtillot, V. "What Caused the Mass Extinction?" *Scientific American*, October 1990.

Benningfield, D. "Where Do Comets Come From?" *Astronomy*, September 1990.

Boss, A. "The Genesis of Binary Stars." *Astronomy*, June 1991.

Dietz, R. "Demise of the Dinosaurs: A Mystery Solved?" *Astronomy*, July 1991.

Verschuur, G. "The End of Civilization?" *Astronomy*, September 1991.

A Glimpse of Life in the Ice Age

I heard a loud thud from below. It must have been a small meteor striking the spacecraft's fuselage, I reasoned. It was a little unusual in this part of space, but no big deal. Then another thud rocked the ship, considerably stronger than the first. I looked out the window. There was a massive vortex of flying rocks rotating around a common center. Before I could react, it was on me. I stared straight ahead into a veil of complete blackness.

The ship was sucked right into the thing like a speck of dust into a vacuum cleaner. The walls of the ship creaked as it rocked sideways and tipped onto its side. Then the ship lurched forward a second time. All seemed calm. I looked around and saw a starry sky.

I found Earth, or what seemed like Earth, below me and a little to the left. I oriented the ship to point toward it.

It was different. At first, I did not recognize it as the Earth. The continents were different. But it must be the Earth, I told myself. What else could it be?

I was somewhere over the Atlantic Ocean. To my right was Europe, where it was evening. White solid ice covered the area where Scandinavia should have been. The Mediterranean Sea had roughly the same coastline I was familiar with, except Italy and Africa were almost joined. A large protuberance stuck out from the northeast part of the continent where the British Isles should have been. To the left of Europe was a white splotch of ice corresponding to Iceland, then a massive area of white ice that ran from the bottom of Greenland all the way to the North Pole.

But nothing could have prepared me for the sight of North America on my left. The peninsula of Florida was swollen, but basically the same shape. Mexico and South America were also swollen, but had shapes that were roughly recognizable. That's how I came to recognize North America. A massive block of ice that came from the North Pole covered the entire area of Canada and the northern half of the United States. There were no Great Lakes. The central part of the United States was dense forest land, not plains. A large lake covered the areas of Utah, Nevada, and Idaho in the western part of the United States. Another lake covered most of southern California in the area of

Death Valley.

Farther to my left lay Alaska, covered by ice in the north and in the south, but green in the middle. It was morning there, and I could make out some shadows of the ice-covered mountains on the central plains. The west coast of Alaska was not there. It simply merged with Asia. I couldn't see beyond that. Throughout this new world, I did not see extensive cloud cover, except in the tropics. On the whole, the world was drier and colder than the one that I had just left. It must have been that thing that the ship got sucked through. It had taken me back in time. Now I was looking at the Earth of 15,000 years ago. I had gone back in time. I was looking at Earth as it existed during the last great Ice Age.

I steered the ship in the general direction of Alaska. I don't know why, but the place fascinated me, and the idea of visiting it during the Ice Age seemed irresistible. I re-entered the Earth's atmosphere and flew in a northwesterly direction over the midsection of the United States. I crossed the snow-covered Rocky Mountains, then flew over the northern part of the large western lake. I never saw the Pacific coastline because of heavy fog.

I continued my descent. The fog broke up as I approached the south coast of Alaska. A solid wall of blue-white ice rose abruptly from a clear blue sea. The ice rose steeply to form rugged mountainous shapes over the land.

I was about 30,000 feet up. As I crossed the coastline, the air become bumpy. The ship lurched about violently. It was probably the worst turbulence I had ever experienced. I struggled to maintain control of the ship as it jerked up and down and from side to side. Below me was nothing but shiny ice. A yawning valley opened up, and then another range of mountains, the Alaska Range. To my left was the large, dominant peak of Mt. McKinley.

The ride got bumpier still as I flew over the rugged ice that covered the Alaska Range. A sudden blast of wind lifted the ship's nose sharply up, then brought it down hard with a crashing jolt that jarred me almost senseless.

Beyond the Alaska Range was an area of shrubs and treeless tundra. The ride continued to be very bumpy. After a few minutes, the tundra gave way to gently rolling hills, covered with green grass dotted with a few trees and shrubs. Many small streams threaded their ways through the valleys.

The turbulence had subsided. I steered the ship toward a flat treeless area and began my final descent. I reached for the ship's communicator. I pretended that I was the pilot of a large passenger jet, and spoke into the microphone. Why not? I might as well have a little fun over this thing.

"Well, folks," I drawled. "Some of you may have noticed some unusual things during the last hour or so. There's nothing to be alarmed about, however. The plane has passed through a singularity in space-time, specifically, a rotating black hole. We were able to successfully negotiate the black hole, and are now flying over central Alaska.

"The bright blue river that you see off to the sides of the aircraft is the Tanana River, and about thirty miles off to the right is the approximate future location of the city of Fairbanks.

"As for the local time—well, from the general position of the sun, it appears to be mid- to late June, and about nine-thirty in the morning, give or take a few minutes. As for the year, it's about 15,000 B.C., give or take a thousand years.

"As you can see, it's a beautiful sunny day in central Alaska. The local forecast calls for partly sunny skies throughout the day with a high temperature in the mid-sixties, and a slight chance of some light showers this evening. The winds are out of the east at twenty-three miles per hour with some higher gusts, and the present temperature is fifty-six degrees.

"We would like to take this opportunity to thank you for flying with us, and we do hope you enjoy your stay in the late Pleistocene. At this time we ask you to fasten your seat belts. We will be landing shortly."

I guided the ship down to a bumpy landing in a flat grassy area. The ship rolled to a stop. I was immensely grateful for one thing. NASA had deemed the environmental conditions of the ship safe enough so that I didn't have to wear a space suit. So I was wearing a pair of running shoes, blue jeans, and a navy blue t-shirt. Somehow, I couldn't imagine stepping out into the world of the Ice Age dressed in a bulky space suit. I opened the door of the ship and slowly stepped out and onto the soft grassy ground of Ice-Age Alaska.

I stood on a flat grassy plain, with snowy mountains in the distance in almost every direction. Small isolated groves of trees stood in scattered clumps. The wind was brisk but not overpowering. The temperature was comfortable. Patches of white, puffy clouds floated through the sky, driven by the steady wind.

I felt something on the back of my neck. I reached back with my hand and felt a large mosquito there. I brushed it away.

"Well, here I am in the Ice Age. What am I going to do now?" I said to myself. I had enough food in the ship to last for a while, maybe a few weeks. Eventually, I was going to have to find my way out of here or figure out how to catch fish or something. I swatted a mosquito on my arm, leaving a small splotch of red blood. Yuk, I hated fish. But what else was there?

I noticed a group of guys walking toward me. They were wearing light clothing that looked like it was sewn together from animal skins.

They had long hair but no beards. I swatted another mosquito on my arm. The men were carrying spears with wooden shafts and pointed stone tips. I waved at a mosquito that buzzed in front of my nose. I held up my right hand with palm facing outward. "I come as a friend," I said.

"Hey, far out," one of them called out. "You're our friend. That means you can help us hunt the mammoths." (For readability, I have translated my conversation with the locals into colloquial English.)

"Hunt the mammoths?" I swatted another mosquito, this time on my wrist.

"Well, what do you think we should do?" another one of them said. "Sit around the campfire and talk about the weather?"

"Well, now that you mention it, that's an idea." I grabbed at a mosquito that was flying by my ear.

"C'mon, man, cut me a break. You can talk about the weather when we're done hunting the mammoths. There ain't nothing you can do about the weather anyways, 'cept talk about it."

I smashed a mosquito hard against my forehead. My palm was spotted with bloodstains. Why weren't the mosquitoes all over these guys? Why me?

"All right, all right," I said. "I'm your friend, and we'll go and hunt mammoths. But since I'm your friend, can't you do something to keep these mosquitoes off me?"

One of them grabbed a small sewn bag that was hanging off his trousers. He handed it to me.

"Use this," he said. I opened the bag, which was filled with syrupy fluid, and spread the fluid on my arms and face. It stung a little but I could deal with that. Those mosquitoes were driving me nuts. I handed the bag back to the guy. I stood for a few seconds. The mosquitoes weren't bothering me anymore.

"Here you go," one of them said, throwing me an extra spear. I caught it. It felt light and flimsy.

"Over there," he said, pointing into the horizon. I stared in the general direction.

"Where?" I said. "I don't see anything."

"See the little clump of tall trees over there?"

"Yeah," I said, squinting. "I think so." They must have been two miles away.

"Now look just to the right."

"Yeah, I'm looking. What's out there?"

"The mammoth herd. There's probably about a dozen of them. You can see the little black dots. They've just gone up a hill or something. That's how we can see them this far away."

"I don't see anything," I said. Jeez, was my eyesight really that bad? How could these guys see mammoths way out there?

"You can smell them, too, a little bit," one of them said.

I tried to smell. I wasn't sure, but maybe there was a faint aroma of pine needles or pollen or something. But nothing like animals.

"We need to plot an intercept course," one of them said. "How fast do you think they're going?"

"They're just sort of ambling along," another said thoughtfully. "I'd say about five miles per hour. They're headed straight to the right."

"And it looks like they're about five miles away. Say we can run at eight miles per hour. Which way should we go?"

During the last several years I had done a little lunchtime running in fits and starts. I was thankful that this happened to be one of those times when I was in decent running shape. I decided to give them some good-hearted teasing.

"Eight miles an hour, eh? So you guys think you can run a 7:30 mile pace over hilly ground, into the wind, carrying spears, and wearing those heavy boots with no arch in the foot, made out of caribou skin? Yeah, right, tell me about it."

"Hey, easy, man. I'm trying to figure out the intercept course we need to take."

I took a minute to work it out. "Here you go," I said, pointing at about a thirty-eight degree angle to the right of the herd. "There's your intercept course." (See Figure 5-1.)

"Yeah, how'd you get it so fast?"

"Well, it was actually quite elementary," I bragged, glad to have the upper hand in something finally. "It was just a matter of taking the arcsine of five-eighths. And by the way, we'll have to run for about forty-eight minutes and cover 6.4 miles total."

"Arcsine, eh? Yeah, right, tell me about it."

I should have known. Trigonometry would not come for another several thousand years.

"No, honest," I said. "I wrote a book on this kind of stuff." Oops. The printing press was not invented until the fifteenth century. They were staring at me doubtfully.

"Wait," one of them said. "He must be right. If we have to go 6.4 miles, and the herd is 5 miles away right now, that's close to a four-to-five ratio. Then we can use the 3-4-5 triangle thing. You know, like I was showing you with the rocks on the ground. It's just about that direction. I think he's right." The group looked at me respectfully. I breathed a little easier.

"So are we going to start?" I said. "Or do you guys just want to sit around the campfire and talk about the weather?"

We started running. It seemed like an easy loping gait. I was immensely thankful that I was wearing my running shoes. I couldn't imagine how these guys could run in those big, cloddy boots, but

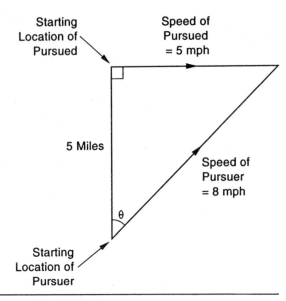

FIGURE 5-1. *Plotting the course of a pursuer. In the case, the pursued is moving at right angles to the line of sight of the pursuer. The pursuit angle for intercept is the arcsine of the ratio of speeds, in this case 5/8. The angle thus computed can then be used to find the total distance the pursuer must cover.*

somehow they were doing it. One of them even looked like he was in his sixties. We jumped over a small stream. The spear was a real nuisance. I was worried about dropping it, or swinging it and hitting someone, or poking myself in the eye. I had to carry it pointed up so it wouldn't swing around too much.

I was feeling warmer after a while, and could feel myself start to sweat. I tried to keep from breathing too noisily. I didn't want these guys to think I couldn't keep up. Everyone seemed to be running so easily. We came to another stream. I took one step into it. My left foot smacked the water, coming down on a pebbly bottom about six inches deep. It was freezing and gave me a real shock.

I could barely make out the forms of the mammoths now. I could see the black dots on the horizon that everyone else saw a few miles ago. I felt good at seeing that. It meant the end of this run. What started out seeming like an easy loping gait was rapidly becoming a painful ordeal. I wasn't sure what was going to happen when we caught up to the mammoth herd. Right now, the most important thing was to get finished with this run.

A puffy cloud blew over to block the sun. It felt refreshingly cool to run in the shade. My spear, which had once felt so weak and flimsy, was getting heavier. I switched it to my other hand. I almost dropped it. My palms were awfully sweaty. I heard the breathing of a couple other guys. That made me feel better. Somebody else was getting tired, too.

The sun came back out from behind the cloud. I could see the big lumbering forms of the mammoths, plodding slowly along from left to right. We were running at an angle to them, aiming for the intercept point. It looked like they didn't see us. I guess we looked pretty small to them at this distance.

I was breathing hard and sweating heavily. My t-shirt had become soaking wet. My legs ached. They were tight. Each step was painful.

"I need to rest," I said to the guy next to me.

"Hey, man, I'm beat myself," he panted. "But we can't rest. We're on an intercept course. If we stop and rest, the mammoths will just walk away from us. Then we'll all go hungry for another day."

Excellent point. I was not running to get exercise or burn off fat. I was running because basic survival depended on it. Those words, coming from someone who was just as tired as I was, gave me the strength of will that I needed.

I could smell the mammoths now. The wind that was blowing from them to us, which disguised our scent from them, clearly transmitted their scent to us. The smell quickly became overpowering and sickening. I tried plugging my nose, but the action was too awkward, since I was already carrying a spear. We must have been less than a mile away now. I could see their shaggy shapes and the long curved tusks of the males. I counted fourteen of the mammoths in all.

"It's time to split up," one of the leaders said. "They're going to be able to see us pretty soon." He motioned to a group of us that included me. "You guys take the rear. We'll go around to the front. It looks like our best shot is at that big bull third from the left. We'll do a 4-4-3 stunt right. All right, let's go."

We split up into the two groups, running at the same pace all the time. My group had angled off to the left to attack from the rear.

"Ever play weak safety?" one of them asked me.

"Weak safety? Huh? Like in football?"

"We're going to line up in three rings—an innermost ring, a central ring, and an outermost ring," he said. He was breathing hard as he ran, and his words were coming out in broken gasps. "You're in the outermost ring. It's an easy job. Just keep your eye on the mammoth and move to intercept him if he tries to break out of the two inner rings."

"Okay," I said. "I think I see."

We were running behind a row of trees. The leader of our group was careful to keep us hidden as long as possible. The mammoths were very close. They were right there in front of us, maybe a few hundred yards away. I could see the grass and the dust that they kicked up as they lumbered by. I could hear their soft grunts as they nudged each other. I could see their hairy tails flicking the mosquitoes away. I kept my eye on our quarry. He looked pretty big. I couldn't understand why we were trying to fight an adult male and not a smaller female or calf. But I was too out of breath to ask about it and get into a conversation.

We circled around to the rear. We were almost directly behind the group of mammoths now. One of them grunted loudly. The others looked over. They saw us.

"Okay, now!" our leader said. "We have to go now."

The group started sprinting in toward the animals and fanning out. The group of mammoths started to trot away, then saw their way blocked by the second group of men, charging now from the front. They stood for a few seconds in a state of indecision, and this gave us the time to complete our encirclement.

If the mammoths now had acted as a group, they could have easily stampeded in unison right through our lines, possibly sustaining only minor injuries from glancing blows of spears. But they didn't. They moved about in various isolated groups. Our quarry, the large bull, was alone in the center of our circle of men.

The four linemen of the inner circle charged him together. As they approached, the bull swung his huge curved tusks back and forth, catching one of the men in the ribs. He fell to the ground and lay motionless. Two of the men threw their spears, hitting the beast in the head and ear. He grunted, turned around, and faced the second wave of attackers.

One of the men charged up to the beast from the front, slid below the animal's front feet, and jabbed the spear into his underside. It was a bold gamble. The animal screamed in pain, its trumpet-like voice booming across the land. The man tried to get up, but the animal stepped on his legs, breaking both of them instantly. The man cried out in pain. I recognized the man. He was the one who knew about the 3-4-5 triangle.

The beast turned again, facing in my general direction, and tried to make his escape. He ran with a slight limp, angling a little to my right. I ran to intercept him head-on.

He saw me, and realized that he was hurt and would probably not be able to get by me with quick maneuvering. He faced me straight on. With an angry screech, he charged directly at me.

"Hold your ground!" one of the men yelled to me. I could see them running over from their positions toward the animal.

Hold my ground? I was about to get killed. Well, hold on. Maybe I wouldn't be killed, I told myself. I was certainly going to get hurt. Maybe I would get some broken bones, cracked ribs, or a concussion.

I held my spear ready to throw it. I could feel my arm shake. I saw the animal's big sorrowful eyes, and the last thing I wanted to do was hurt this magnificent beast that was headed one day for extinction. But I had to. I had to because the survival of my small group of friends depended on it. I had to because others had already risked their lives and had been seriously injured in pursuit of this goal. I focused on the animal's heart. I would try to plunge the spear between its tusks, avoiding its large trunk, and hitting it in the heart.

The mammoth came at me. It was just a few feet away. It gave out a loud, angry roar. I kept my eye on the skin over its heart, and threw the spear.

It was about then that I woke up. It was a heck of a dream, and I still spend time wondering how things turned out. Well, it wasn't actually a dream—not a "dream" dream—just sort of a daydream. Oh, okay, I admit it, I made up the whole thing. I've never been an astronaut, and I don't know how to fly. But I guess if you know your geography, you could tell I didn't know what I was doing when I landed the plane in that tricky crosswind.

But we're not going to close the book yet. Despite the smattering of untruths, there is plenty of interesting science that can be dredged out of the story.

A visitor from space viewing Earth at the time of the Ice Age might not recognize it as the same Earth of modern times. The continents were shaped far differently than they are today.

This is not because of the effects of plate tectonics and continental drift. Continents move so slowly that significant positional changes can only be seen on a time scale of millions of years. My dream took place about 17,000 years ago, not long enough to notice any appreciable effects of continental movement.

The reason that the continents were shaped differently was because of the redistribution of water from the oceans to the glaciers. The Earth is basically a closed system. Water does not leave the Earth and its atmosphere and oceans, nor does it get produced there. The total amount of water in the system is roughly constant. Therefore, the water that was used to create the massive glaciers had to come from somewhere. It came from the oceans. Ocean levels were about 400 feet lower than they are today.

Most continents assumed larger, swollen shapes. Land that was usually submerged by water appeared above the surface. This effect created several interesting and dramatic "land bridges."

England and Ireland were linked to each other and to the European continent. Australia was linked to Tasmania and New Guinea. Japan was linked to South Korea and China in the south, and to Sakhalin Island and Siberia in the north. The islands of Japan were linked together also, and the present day Sea of Japan was then an inland lake.

Italy and Sicily were linked, and the two came very close to touching the coast of North Africa. The Philippine Islands were linked to each other and to Borneo, which was linked to Southeast Asia. Newfoundland was linked to Canada.

Perhaps the most dramatic land bridge of all, however, was the link that was created between Alaska and Siberia. (See Figure 5-2.) This link was not a "bridge" in the standard sense, which implies a narrow strip of land. It was about 600 miles across from north to south.

Scientists today agree that the first humans to populate North America were immigrants from Asia. The strongest argument to support this comes from analyzing the dental configurations of the two peoples. One peculiar type of incisor tooth is only found in the remains of American Indians and aborigines of Asia. Skeletal remains of early Europeans and Africans do not show this peculiar tooth formation.

FIGURE 5-2. *The Ice Age world as it might be viewed from outer space by an astronaut.*

Whether the first Americans arrived by land over the land bridge or by sea in primitive canoes is a topic of current debate. But the majority opinion seems to be that they came by land. An archaeological site called Bluefish Caves in northwest Yukon Territory proves that humans lived there at the same time that the land bridge existed.

During the height of the last Ice Age, glaciers covered practically all of Canada and a good portion of the continental United States, including the entire Great Lakes region and the present-day New York City metropolitan area. Glaciers invaded Alaska from the Alaska Range southward to the coast, and from the Brooks Range in the north, northward to the Arctic Ocean. Central Alaska and the land bridge linking Alaska to Asia were spared from the ice. Large amounts of precipitation were necessary for glaciers to advance, and central Alaska at that time did not receive enough winter snow to support an advancing glacier.

Most of the world was drier during the Ice Age than it is today. The oceans were colder. Evaporation, which is a process that requires heat, did not occur as easily in the oceans of the Ice Age.

An interesting exception to this rule was the American southwest. Today, this area lies under persistent high pressure in an area of the world known as the *horse latitudes*. The horse latitudes are an area of almost no wind, bounded on the north by the prevailing westerlies and on the south by the tradewinds. But because of the presence of the enormous glaciers covering Canada, storm tracks were pushed southward over North America, and the region of the horse latitudes was squeezed into Mexico.

Although certainly not a densely forested region, the American Southwest then received enough precipitation to support the huge Lake Bonneville, of which the Great Salt Lake of Utah is a mere tiny remnant. Another lake filled the area now occupied by Death Valley, California.

Ice-age temperatures worldwide averaged about six degrees Fahrenheit colder than today, but most of the effect was felt in the higher latitudes. Part of this is because the large masses of glacial ice have a high albedo; that is, they reflect most solar radiation that strikes them. The high albedo of the ice creates a sort of "feedback loop." As temperatures cool, more glaciers form, the albedo is raised, temperatures cool even more, and so on. There was probably little temperature change noticed in the tropics during the Ice Age. But temperatures in the continental United States and in Europe were probably about twelve degrees cooler than they are today. Regions near the large ice sheets were particular sufferers of the effects of lower albedo, and hence less heating by the sun. It has been suggested that temperatures in the American Midwest may have been as much as eighteen degrees colder than present, because of the close proximity to the enormous Canadian ice sheets.

Because of the land bridge between Alaska and Asia, central Alaska had more of a continental climate than it does today. It was not as surrounded by ocean as it is today. This effect tended to warm the summers and cool the winters. But central Alaska was also close by ice sheets in several directions. There was the large Canadian ice sheet to the east, with tongues that probed southern Alaska and northern Alaska. This nearness to the ice would tend to cool the climate.

Likely ice-age temperatures in central Alaska, taking these effects into account, probably averaged about five degrees cooler than today in the summer and fifteen to twenty degrees cooler in the winter. Temperatures of forty below zero during the long periods of winter darkness were probably quite common.

The Tanana River, or something like it, almost certainly flowed through central Alaska. Fed by melting snows from the great ice sheets, it was particularly impressive during summer. But it certainly did not empty into the Bering Sea, which did not exist at the time. The river probably continued into the land bridge area, eventually emptying in the Pacific to the south.

Alaskan weather was dominated by persistent high pressure over ice sheets to the north and east. Local winds, as a result of clockwise rotation about a high pressure cell (to be discussed in Chapter 6) came mainly from the south and east. Southerly winds, probably the more usual case, caused frequent snow along the coast.

Strong differences in temperature led to strong winds. The winds at the edges of the ice sheets were probably extremely strong, perhaps in excess of 100 miles per hour. An airplane flying through such conditions would experience violent turbulence.

Windy conditions prevented the growth of trees. Ice masses were surrounded by strips of tundra, perhaps fifty miles wide, where trees could not grow. Farther from the ice masses, some trees could grow, along with various other grasses. But some amount of wind prevailed even at long distances from the ice. Central Alaska during the Ice Age was a very windy place at all times of the year.

Fossils of ice-age animals found in central Alaska suggest that a large population of mammoths, horses, and bison lived there. Since all these animals were grass-eaters, scientists have proposed that central Alaska was an area of grassland. Arctic tundra could not have provided enough grazing land to support this type of animal population.

The mammoth provided necessary food and clothing for the people living in the Ice Age. Mammoth hunting was therefore a vital skill to them to ensure their survival. Methods of hunting mammoths varied throughout the ice-age world. Where possible, hunters tried to make use of the terrain to hunt. Hunters would often try to herd the animals off a cliff, or trap them in a steep canyon.

The most challenging hunting of all, of course, occurred in regions such as central Alaska, where the terrain consisted mostly of rolling hills and flat, grassy plains. Mammoths very seldom ventured into Alaska's mountains, which were usually glaciated.

There can be little doubt that hunters in this area employed ingenious methods of cornering their quarry. Large, ornery bulls may have been singled out on the theory that others in the herd might not come to its defense. Ice-age people may have had sharper senses of sight, smell, and hearing than modern people. Their survival in the environment certainly depended on their abilities to locate herds of game at a long distance. The senses would be extremely important, unless by good fortune they happened across tracks left by the animals.

Ice Agers were also in excellent athletic condition compared to modern people. Again, their survival depended on their ability to run long distances to track and hunt prey over a large area.

Ice Agers doubtless had a good intuitive sense of geometry, as this was necessary in tracking game. But they certainly had no formal theory to guide their thinking. (The invention of formalized geometry is usually credited to the ancient Greeks.)

A particular problem that must have been faced by the ice-age inhabitants of central Alaska was the profusion of mosquitoes. The climate at that time was cooler and drier than today. Therefore, the mosquito problem was not as bad as in modern times. Nonetheless, melting snows during spring and summer created areas of standing water where mosquitoes could breed. Although not as bad as today, the mosquito problem almost certainly existed during the Ice Age. Solving it somehow was necessary for mankind to function efficiently in the area.

Mosquitoes are attracted to human hosts by body moisture, lactic acid, and heat. A navy blue t-shirt, or any such dark color, is a poor reflector of radiation, and thus absorbs heat, attracting mosquitoes.

Mosquito repellents are chemicals that affect the mosquitoes' somewhat automatic behavioral response. A large class of repellents, known as pheromones, is actually used by insects for the transmission of sexual communication signals. Used in small doses, pheromones attract mosquitoes, but in large doses they are effective repellents. Hence the use of pheromones is a little tricky. Natural repellents are more effective, since they can repel mosquitoes when applied with any level of dosage.

Some natural repellents exist as chemicals in the fluids of several plant species. One such repellent may have existed in the fluid of the balsam poplar tree, which grew throughout ice-free Alaska during the Ice Age, as confirmed by fossil evidence. The chemical benzyl benzoate, which exists in the fluids of most balsam trees, is still used today as an effective insect repellent.

My personal view of Ice Agers is of a friendly, humorous, fun-loving, and witty people. There is certainly no science to support this view, and it differs from the popular fiction that has been written about this era. But this is the only way that I can imagine it.

Consider that the life of the Ice Ager was very stressful. He was worried about getting food on an almost daily basis. He needed to store enough food to last through a long and bitterly cold winter. He did not have television, video games, or other modern fun-offerings. One of his only pleasures was the joy of friendship and conversation with the other members of his clan. It stands to reason that his conversation sparkled with good-humored wit and teasing.

In the case of Ice Agers in Alaska, the long winter months were spent mostly in darkness huddled around the fire. Conversations and storytelling were probably key to passing this time.

Ice Agers probably did not have formal names in the sense that we have today. People were called by nicknames that sometimes stuck and sometimes changed to something else. In my story, I may have been called something like "he who wears a dark t-shirt in mosquito country."

The woolly mammoth was roughly the same size as the modern day African elephant. It stood about ten to thirteen feet at the shoulders. The tusks of the male were enormous, often growing to lengths of fifteen feet or more. The tusks curved inward, an interesting evolutionary adaptation that worked to the mammoth's advantage in digging for food through snow. The curved tusks could also be used for defense against predators, but straight tusks probably would have made more effective weapons.

An enticing ice-age mystery is the cause of the extinction of the magnificent woolly mammoth about 10,000 years ago, as the Ice Age was coming to a close. In fact, the mammoth was just one of several animal species that all became extinct at about that time. Affected were larger animals, such as the woolly rhinoceros, the saber-toothed tiger, the cave bear, and the North American giant ground sloth.

One theory for the extinctions is overhunting by human beings. A major argument against this theory, however, is that many of the affected animals were not hunted by man. Another argument against this is the coincidental simultaneity of many extinctions.

A more plausible idea is the changing vegetation patterns across the world as the Ice Age waned. The waning of the Ice Age occurred quite rapidly, as opposed to its onset, which was gradual. Vegetation changes at the end of the Ice Age were so rapid that species were unable to adapt in time. In Alaska, for example, the open grassy steppe of the Ice Age changed quite rapidly to forests, woodlands, and tundra as the ice receded. Animals such as the mammoth, which survived by grazing, could not adapt to the new vegetation, and thus could no longer survive in Alaska. Large animals would be the most dependent on the vegetation for survival, and hence the most likely to become extinct.

Still other theories involve volcanic eruptions or meteor impacts. In my opinion, these theories fail to explain the selectivity of the extinctions. Furthermore, they are not backed up by geologic evidence. At this point, in my opinion, the most believable theory is the changing vegetation patterns.

The Ice Age presented hardships for human beings in all parts of the world, not just in the Arctic areas. As the amount of livable land decreased, the competition for it became fiercer. Human beings in the tropics were thus also affected by the great ice sheets that lay thousands of miles away.

Perhaps one reason that the Ice Age is such an interesting subject to most of us is that it was, in a sense, mankind's finest hour. It was a time of intense competition for survival. Human beings competed with the likes of the fearsome saber-toothed cat. Their ability to survive was definitely put to the test, and extinction was a very real possibility. But they competed, and, somehow, miraculously, they won.

Human beings won by surviving, and by surviving they made it possible that we might also live. In the process, they accomplished feats of courage that we of today can scarcely imagine. Armed with puny weapons of wood and stone, they fought elephants.

BIBLIOGRAPHY

Books

Chorlton, W. *Ice Ages*. Alexandria, VA: Time-Life Books, 1988.

Fagan, B. *The Great Journey: The Peopling of Ancient America*. London: Thames and Hudson, 1987.

Laughlin, W., Marsh, G., and Harper, A. *The First Americans: Origins, Affinities, and Adaptations*. New York: Gustav Fisher, 1979.

Lucas, J. *Becoming a Mental Math Wizard*. White Hall, VA: Shoe Tree Press, 1991.

Michener, J. *Alaska*. New York: Random House, 1988.

Pielou, E. *After the Ice Age: The Return of Life to Glaciated North America*. Chicago: University of Chicago Press, 1991.

Periodicals

Garrett, W., editor. "Where Did We Come From?" *National Geo graphic,* October 1988.

Mehringer, P. "Weapons of Ancient Americans." *National Geographic*, October 1988.

Putman, J. "The Search for Modern Humans." *National Geographic*, October 1988.

What Causes Ice Ages?

Earth is about 4.5 billion years old. For most of that time, our planet has been a hot, dry, inhospitable place. Mankind has evolved and developed completely during a prolonged cold spell in the Earth's see-sawing climatic picture.

The cold spells have all been marked by glacial ice. Today our glacial ice is mostly in Greenland and Antarctica. Humans are not forced to live in or near these areas. But the presence of the glacial ice marks the current period in the Earth's history as a cold one. Geologists use several terms to describe the Earth's alternating hot and cold spells. There are ice eras, ice epochs, ice ages, glaciations, and interglacials. We are presently living in a period known as the Holocene interglacial. The Holocene was preceded by a period known as the Pleistocene ice epoch. The Pleistocene began about 2.4 million years ago, and lasted until the end of the most recent ice age, when the glaciers started their retreat. The Pleistocene is just a small part of a larger ice era, which began about sixty-five million years ago and includes the present.

The terminology can be a bit confusing. A simplification that approximates history might be that warm spells alternate with cold spells, and that cold spells are subdivided into ice ages and interglacials. Cold spells are brought on by the relative positioning of the continents on the Earth's surface. Because of plate tectonics and continental drift, continents are always shifting their positions. This process, of course, is extremely slow and requires millions of years for noticeable changes to occur.

During cold spells, continents are placed so as to restrict the amount of heat exchange between tropical areas and polar areas. At the present, the continents are positioned in such a configuration. The South Pole is covered by a continental land mass, Antarctica. The temperature of the interior of Antarctica cannot be moderated by nearness to ocean water. Thus, the flow of heat from the tropics to the South Pole is impeded. A similar situation exists at the North Pole. The North Pole is completely surrounded by land masses, thus preventing the free circulation of ocean water from the tropics.

Earth has much more water than land. Therefore, a configuration of continents such as the present one is unusual. During most of the Earth's history, tropical waters have been able to find pathways to the polar areas. Thus, the formation of glaciers in the Earth's polar regions is a relatively rare event.

The cold spells can also be explained in terms of the Earth's heat budget. In simple terms, the Earth and its atmosphere warm up or cool down based on the amount of heat absorbed from the sun versus the amount of heat reflected or radiated into space. If the amount absorbed exceeds the amount reflected or radiated, the Earth warms. If the reverse is true, the Earth cools.

Tropical regions absorb more solar radiation than they reflect or radiate. The reverse is true of the polar regions, and it is only at about forty degrees of latitude that an approximately balanced heat budget exists.

If there were no oceans, the polar regions would get constantly colder and the tropical regions would get constantly hotter. The oceans circulate heat from the tropical regions to the polar regions. This effect can actually be quite dramatic. The average January temperature in Reykjavik, Iceland is much warmer than that of Omaha, Nebraska. The average July temperature of Yakutsk, Siberia is warmer than that of San Francisco.

A change in the heat budget of one part of the world can affect the overall climate. For example, if the polar regions are allowed to cool, ice can form there. The ice will then reflect much more of the sun's radiation, and the polar regions will cool even more. Some of the cold will circulate by winds and ocean currents throughout the rest of the world.

Whereas the configuration of the continents is the commonly accepted explanation for Earth's long-term cold spells, the ice ages themselves are harder to explain. Earth has been in a cold spell now for sixty-five million years. During all that time, there have been glaciers on the Earth. But the extent of the glaciers has varied dramatically.

The advances and retreats of glacial ice have not always been gradual. During a period of time known as the Little Ice Age, climatic change occurred rapidly and noticeably over the span of a few generations. The Little Ice Age began about 1350 and continued to about 1870. Before the Little Ice Age, there were Nordic settlements on Greenland, which was literally green. Cattle were bred on the stretches of grassland. A worsening of the climate began about 1350. Agriculture and livestock breeding were drastically affected. In the space of under 100 years, most Greenlanders left.

On the European continent during the same time period, glaciers advanced into Alpine valleys and snow lines on mountains were noticeably lower. The growing season was shorter by about three weeks, and crops could only be grown at lower elevations. The effect was severe on large parts of the population, particularly those who were living at or near the subsistence level.

Signs of advancing glaciers during this time can be found in the Rocky Mountains, where many trees were left damaged by the advancing ice. Evidence points to glacial advance in the Arctic as well, and on islands in the North Atlantic.

There is some evidence that the amount of solar radiation reaching Earth declined during the period of the Little Ice Age. Astronomers of the time noticed a curious absence of sunspots, and the aurora borealis was only rarely visible in the polar regions. Tree rings showed heavy concentrations of radioactive carbon-14, an effect associated with decreased solar radiation.

A warming trend followed the Little Ice Age, continuing into the 1960s. Then the climate cooled for the next two decades. Warming in the 1980s has often been attributed to the greenhouse effect, but this is questionable in the context of Earth's long-term climate history.

There remains the central question of this chapter. What causes the great ice ages? What makes the glaciers advance as far south as New York City, then retreat all the way to the northern islands of Canada? To get a handle on this, we need to start with some more basic questions. What makes precipitation fall to Earth? Why do glaciers form and what makes them move?

THE BASICS OF WEATHER

Molecules of warm air move faster than molecules of cold air. For this reason, the molecules of warm air are able to spread out more. Warm air becomes less dense than cold air, whose molecules, by their lack of rapid movement, are more compacted together. When warm air and cold air mix, Archimedes' Principle takes effect. A buoyancy force acts on the lighter warm air, forcing it to rise over the cold air. And thence comes the first law that governs the formation of the weather: Warm air rises, cold air sinks. It is a simple rule with widespread implications.

How Precipitation Forms

Weather is formed in the part of the atmosphere at the lowest altitude, the troposphere. In the troposphere, temperature drops with altitude. The troposphere receives most of its warmth from heat radiated from the Earth. Therefore, it is reasonable that its temperature goes down with increasing altitude.

Above the troposphere is the stratosphere, where temperature increases with increasing altitude. In this region, radiation from the sun has a greater effect than radiation from the Earth, thus the reversal of the trend.

Weather is formed in the troposphere, not the stratosphere, because of the relationship between temperature and altitude in the troposphere. Warm air rises in the troposphere. As it rises, it becomes colder.

This leads us to the second major law of weather: Warm air is capable of holding more moisture than cold air. This is because warm air is less dense than cold air, its molecules are less tightly packed together, and hence there is more room in the warm air for the water molecules. Any mass of air has a saturation point, which is the minimum temperature that the air can have and still be able to hold its moisture.

When warm air rises, it cools until it reaches this saturation point. That is where clouds form in a process called *condensation*. Clouds are made up of tiny water droplets or ice crystals, the result of the condensation of water vapor.

Van der Waals forces create rain or snow. Tiny water droplets or ice crystals in clouds are so light that they do not fall to the ground. They are whipped around by air currents and remain aloft in the cloud. Because of van der Waals forces, tiny droplets or crystals that collide with each other tend to stick together. Eventually, droplets or crystals grow through many collisions to a large enough size where they fall to the ground as rain or snow. The process is often aided by the presence of aerosols such as minute dust particles that can act as seeds for droplets or ice crystals to coalesce.

Evaporation

Evaporation is the process that changes liquid water to a vapor. The process requires heat to break the attractive forces of the liquid that hold its molecules together. The heat can be supplied either by the evaporating substance (such as warm air) or by a drop in temperature of the liquid.

The rate of evaporation also depends on the relative humidity of the evaporator. The drier the air, the faster evaporation can occur, since the dry air is capable of adding much more water vapor to its stores.

A situation that is especially conducive to precipitation is cold air blowing over warm water. The water is warm enough to permit evaporation. The water vapor created by the evaporation is warmer than the surrounding cold air, and rapidly rises, cools, and condenses to form clouds.

High and Low Pressure Cells and Winds

High pressure is created by air that is pressing down on the Earth. Low pressure is created by air that is rising up from Earth. There is some connection between a low pressure cell and stormy weather, since low pressure implies rising currents of warm air, which may lead to precipitation. Areas of high and low pressure exist in bands throughout Earth. They can be understood through a description of the air flow on the Earth.

Our starting point is with the tropical air near Earth's equator. This air, which is always warmed by abundant sunshine, rises and begins to cool. It creates precipitation, and, as you can easily verify, all the areas of the Earth that bracket the equator have rainy climates. Just as the process of evaporation requires heat, the process of condensation (the opposite of evaporation) releases heat. So the air that has just risen over the tropics remains warm. It then begins to circulate north and south over the Earth.

As the air moves farther from the equator, it cools, becoming denser and heavier, eventually reaching a point at which it sinks to the Earth. This "point," which is really a pair of bands that encircle the Earth, one in the north and another in the south, marks the location of the horse latitudes. The horse latitudes are bands of chronic high pressure and very little wind that encircle Earth, and are recognizable as desert areas. In the northern hemisphere, the horse latitudes include the American Southwest, the Sahara Desert, the Arabian Peninsula, parts of Iran, Pakistan, and Afghanistan, and the Taklimakan and Gobi Deserts of China. In the southern hemisphere, the horse latitudes include the Kalahari Desert of Africa, the deserts of Australia, and the Atacama Desert of Chile.

Besides the horse latitudes, the other regions of chronic high pressure are the two poles. This is where the cold dense air presses against the surface of Earth, refusing to rise.

Local winds in an area are determined by the area's location with respect to a high pressure cell. Air moves from areas of high pressure to low pressure, so air will always be flowing outward in all directions from a high pressure cell. The direction of air movement is then bent because of the Earth's rotation by an effect known as the Coriolis force.

The Coriolis force is a key idea in understanding weather patterns. Unfortunately, it is not so easy to comprehend.

Start by imagining a spinning globe. The points on the equator are moving the fastest. The points at the poles are not moving at all. Points in the middle are moving at intermediate speeds. It all depends on the radius of rotation. Points at the equator make one complete circle in the same amount of time as points a little off the equator.

But points on the equator have to make the biggest circle, so they must be moving the fastest. Points on the Earth move slower and slower the farther you go from the equator. Now imagine a high pressure cell—let's say it's over Arizona. What will happen to air as it moves away from Arizona to the north, south, east, and west?

Points to the north are moving slower than points in Arizona, because they are farther from the equator. As air, which started in Arizona and is moving at Arizona's speed, goes to the north (let's say to Utah), it finds itself moving faster than the Utah air. It rotates around the Earth faster than the Utah air. We see it bend to the east as it spins around the Earth at a faster speed than the air around it.

Points to the south are moving around the Earth faster than they are in Arizona. So air moving from Arizona to the south will fall behind the southerly air and bend to the west.

For air moving to the east, we need to invoke the concept of *centrifugal force*. Centrifugal force is traditionally defined as the tendency of a body in circular motion to fly off in a straight line. Another way to interpret centrifugal force, however, is the tendency of a body in circular motion to match speeds with its environment. The second interpretation is useful when working with bodies in rotating frames of reference; for example, a spinning table top or the rotating Earth.

If a body is set spinning on top of a rotating table, it will move outward or inward accordingly, depending on its speed compared to the rotating table area underneath it.

The comparison holds for particles moving on the face of the Earth. Particles moving eastward are moving faster than the Earth below them. They tend to match speeds with their surroundings by bending to the south. Here, air is moving faster, since it is describing larger circles around the Earth. Likewise, air that is moving westward is rotating slower than the Earth beneath it, and curves to the north to match speeds with similarly slow-moving points on the Earth.

I know this is hard to visualize. If you want, you can just remember the rule that in the northern hemisphere, air circulates in a clockwise direction around a high pressure center. In the southern hemisphere, air circulates in a counterclockwise direction around a high pressure center. Air moving from the horse latitudes high toward the equator will bend to the west, whether it is in the northern or southern hemisphere, by the rule of the last paragraph.

This effect accounts for the trade winds that blow through the tropics from east to west. Air moving from the horse latitudes high away from the equator will bend to the east. This effect accounts for the prevailing westerlies of both the northern and southern hemispheres' temperate zones.

The nearest high pressure center can usually be located from the wind direction that you observe. In the northern hemisphere, because of the clockwise rotation of wind around a high pressure cell, this is a simple matter. It is slightly more complex, however, because friction on the ground resists the action of the Coriolis force. The following rule is often given: Stand with your back facing the wind, make a forty-five degree turn to your right, and the high pressure cell will then be to your right. If you are in an airplane, you will not see the effects of ground friction opposing the Coriolis force. Thus, if a high pressure cell is to your east, you will feel wind on the ground coming from the southeast, but the wind aloft will be coming from the south.

Air Masses and Fronts

The air in the atmosphere can be visualized as consisting of large air masses, each of which has its own characteristics. Air masses can be categorized as hot or cold, wet or dry. Examples are continental polar (cold, dry), maritime polar (cold, wet), and maritime tropical (warm, wet).

Fronts are created by colliding air masses of different types. The dynamics of the front are determined by the temperatures and the moisture contents of the two air masses. In all fronts, the collision of air of different temperatures forces the warmer air to rise. Precipitation can result. Since fronts can cause warm air to rise, local highs and lows are created by fronts. A particularly interesting front is the "polar front," which results from the collision of air moving down from the poles with air moving up from the horse latitudes. Most storm systems in temperate regions are the result of clashes of the two air masses at the polar front. The jet stream, a high velocity river of wind that cuts across the temperate zones, marks the jagged and ever-changing boundary of the polar front.

North American Weather During the Ice Age

The start of the Ice Age glaciation probably occurred when summers in northern Canada became too cold to melt the winter snows. As the glaciers grew larger and larger, they became big enough to create a unique chronic high pressure cell. Cold, dense air moved outward from the high pressure cell, and became affected by the Coriolis force. With the high pressure sitting stationary in Canada, prevailing winds over the western Atlantic were out of the north, those over the continental United States were out of the east, and those over Alaska were out of the south.

Conditions for precipitation were enhanced as the cold northerly winds over the Atlantic blew over the warm water of the Gulf Stream. Winds then blowing out of the east over the continental United States brought snow to feed the glaciers.

The horse latitudes were farther south than they are today. Cold air from the glaciers mixed with warm air from the horse latitudes to form a polar front across the central and southern parts of the continental United States.

Since ocean temperatures were cooler than they are today, less evaporation occurred from the oceans. The presence of the polar front in a more southerly location more than compensated by providing the American Southwest with enough rainfall to fill the now dry Lake Bonneville.

The American Midwest during the Ice Age was heavily forested, and also wetter than it is today. Prevailing southerly winds had a warming effect on the climate of Alaska, although it was still a colder place than it is today.

GLACIER FORMATION AND MOVEMENT

Simply stated, a glacier begins to form when snow accumulates in a certain area and does not completely melt during the summer. Snow under pressure becomes ice as air bubbles are squeezed out from between snow crystals. The pressure to change the snow into ice comes from the weight of the snow itself. It is the same effect as the increasing pressure you experience as you go to deeper depths of a swimming pool. You feel the pressure of the water above you.

The ice formed by the process of snowfall and pressure is not a glacier yet. A pile of ice does not make a glacier. A glacier has to move under its own weight. A general rule is that a pile of ice has to reach a thickness of about sixty feet for it to become big enough to move under its own weight.

When a pile of ice reaches this critical thickness and begins to move, there are two ways that it can move. The two ways are termed *internal deformation* and *basal sliding*.

Internal deformation results from the glacier changing shape over time. If you visualize a glacier as a large circular slab of ice, new accumulations of ice in the center of the glacier will change the glacier's shape. It will slowly become dome-shaped instead of pancake-shaped.

The new accumulations will redistribute themselves, being pulled by gravity to lower areas. They will flow over the bottom layers, which may be frozen to the ground and therefore unable to move. But the glacier will still advance by the flow of the upper layers. If left alone for a long time, the glacier will form a new pancake shape, only covering more area than previously.

Basal sliding, the second way that glaciers move, requires a thin layer of water to act as a sliding mechanism between the glacier and the ground. Sometimes the layer of meltwater can be created by heat from the resulting friction as the glacier moves over the ground, causing it to move even faster. Another possible source of the meltwater is melting snow or ice in the top layers of the glacier, which finds its way down to the ground layer through minute cracks.

Glaciers that move by basal sliding are usually formed in mountains and move downhill toward valley areas. Glaciers that move by internal deformation are usually formed in polar regions. Their bottom layers are usually frozen to the ground, so no basal sliding can occur. Basal sliding, assisted by gravity, is the fastest way glaciers can travel.

PIECING TOGETHER THE EARTH'S CLIMATE HISTORY

The Earth contains abundant evidence of its climate history, but that evidence is not easy to get at. Drilling ice cores in Greenland and Antarctica provides one source. Core samples of ooze from the bottom of the ocean are another. Locating fossils of organisms that lived in cold or warm climates is a third.

Once a sample of material is taken for analysis, it needs to be "dated"—its age must be determined somehow. Then it must be analyzed to glean whatever useful information it has to offer. Both of these operations are tricky.

Dating a Sample

The most common dating method involves radioactive elements as described in Chapter 4. The age of the material can be calculated by measuring the ratio of parent to daughter products in the sample.

A second method is to count sections or bands of the material. Just as the age of a tree can be determined by counting its rings, so also can the age of an ice core sample sometimes be determined by minute color changes that represent seasonal freezing and thawing.

A third method is to date the sample based on its position relative to a different particle of a known age. Datable fossils or volcanic dust in the vicinity of the sample can provide valuable clues as to the sample's age.

A fourth method makes use of the phenomenon known as geomagnetic polarity. Earth's magnetic field has flip-flopped several times in its history. The effects of this are, of course, noticeable simultaneously at all points on the Earth.

Rocks containing iron orient themselves in the direction of the Earth's magnetic field at the time of their creation. Undeniable evidence of geomagnetic polarity can be found in the analysis of the ocean floor surrounding the Mid-Atlantic Ridge.

Strips of oppositely magnetized ocean floor appear symmetrically around the ridge. The effect can only be explained through plate tectonics and the seafloor spreading that results at the divergent plate boundary marked by the Mid-Atlantic Ridge. The youngest sediment is found closest to the ridge. As new sediment is created, it pushes the older sediment out, farther away from the ridge as the seafloor spreads. The bands of oppositely magnetized bits of iron on the ocean floor provide the necessary information to determine the history of the magnetic flip-flops.

Some seafloor material can be dated by the modern potassium-argon method. The method is very appropriate for two reasons. First, since potassium radioactively decays into argon with a half-life of about one billion years, the method is accurate over a long period of time. Second, since almost no minerals contain argon as one of their initial constituents, argon makes a perfect daughter product.

Thus, some seafloor material can be accurately dated using the potassium-argon method, then matched with the magnetic orientation of iron compounds in its vicinity. In this way, the history of Earth's magnetic reversals can be determined. The knowledge can then be used to help date other samples anywhere in the world.

Analyzing the Samples

The best method yet for analyzing samples to determine climatic conditions involves two isotopes of oxygen—oxygen-16 and oxygen-18. Oxygen-18 contains two extra neutrons, and is therefore heavier than oxygen-16. The waters of the oceans contain both isotopes. Ice found on land also contains both isotopes.

More heat is required to evaporate water containing oxygen-18 because of its extra mass. This is the key idea. As ocean temperatures fall, less of the water containing oxygen-18 can be evaporated. Then more of it will be in the oceans, and less of it will be on the land. As ocean temperatures rise, more of the oxygen-18 will evaporate, leaving less in the oceans and more on the land.

The key to determining Earth's climate at any given time lies in calculating the ratio of oxygen-16 to oxygen-18 from cores taken from the seafloor, and comparing it to cores taken from land ice corresponding to the same time period. Where is most of the oxygen-18? If it is on the land, the climate was warm. If it is in the sea, the climate was cold. The ratios of oxygen-18 to oxygen-16 on the land and in the sea can be used by scientists to estimate the Earth's average temperature and extent of glaciation throughout history.

Ice core analysis can be used to determine the composition of the Earth's atmosphere through history. The key here is in capturing pockets of air that get trapped in small bubbles of the ice. In this way, not only the chemical composition of the atmosphere can be determined, but also the presence of atmospheric dust that can block the sun's radiation.

Finally, the chemical composition of the ice itself in an ice core can be analyzed. Of special interest are the concentrations of two radioactive isotopes, carbon-14 and beryllium-10. Both of these elements are produced by reactions in the atmosphere that require neutrons from cosmic radiation. In general, the concentrations of the elements carbon-14 and beryllium-10 on the Earth at any one time are proportional to the amount of low-energy cosmic radiation entering the Earth's atmosphere. The low energy radiation contains the neutrons necessary for the creation of these elements.

What this means in terms of climate data is that heavy concentrations of carbon-14 and beryllium-10 correspond to solar minima. Sunspots, a magnetic phenomenon of the sun, are not present during periods of solar minima. Sunspots, because of their magnetic properties, are able to deflect the low energy cosmic radiation. Thus, heavy concentrations of carbon-14 and beryllium-10 in ice samples reflect periods of low solar activity.

A different method to determine the Earth's climate history is in the analysis of fossils. Some species of marine life thrive in warm water, others in cold. An analysis of the concentrations of such organisms' fossils in seafloor cores can, in theory, provide climate information. This method is trickier, however, than the above method of oxygen isotope analysis. Many factors, such as ocean water salt content and food source availability, can also affect the ability of such organisms to survive.

The link between population density of these organisms and water temperature is not so direct. In order for this technique to be useful, more data must be collected. An advanced statistical technique called *factor analysis* must be used on the data to try to isolate the effect of temperature.

Another way to determine climate history is based on the fact that, as glaciers grow on the land, sea levels must necessarily fall. Certain species of coral grow only at known water depths. Dating of fossils from coral at various ocean depths can determine the history of the Earth's sea levels, and therefore the history of its glaciation.

ICE AGE THEORIES

"What causes ice ages?" is an intriguing mystery. It is intriguing because, along with "What killed the dinosaurs?", there are many competing theories. For a theory to be acceptable, four conditions must be met: the theory must be consistent with all known facts; the theory must be able to explain the events; the theory must be complete; and, finally, the theory must be believable as it stands. The last criterion, often called *Occam's Razor*, compels us to accept the simplest of many theories if all meet the first three conditions. Keeping these principles in mind, we will now consider various theories that attempt to explain the ice ages.

Carbon Dioxide

Carbon dioxide is a greenhouse gas. It blocks the heat radiated by the Earth but allows radiation from the sun to pass through. Increases or decreases in the atmospheric concentration of this gas can therefore affect the Earth's heat budget. Analysis of air pockets trapped in ancient ice cores shows that concentrations of carbon dioxide in the atmosphere of the ice ages were much lower than they are today. Could lower levels of carbon dioxide usher in an ice age?

The answer is unclear. Carbon dioxide constitutes an insignificant .03% of Earth's atmosphere, but it is an efficient heat trap. One calculation estimates that the removal of all carbon dioxide from the atmosphere would reduce the average temperature of the Earth by twenty-seven degrees Fahrenheit. The magnitude of this number suggests that a modest decline in carbon dioxide's presence could be sufficient to plunge the Earth into an ice age. Rough calculations based on the amount of carbon dioxide present in air molecules of the ice ages are consistent with the temperature drops in the ice ages. Could it be that simple? Could cyclical rises and falls in the levels of atmospheric carbon dioxide be the cause of ice ages?

The theory explains ice ages and is consistent with the facts. It is also believable. The problem with the theory is that it is incomplete. Why should the level of carbon dioxide in the atmosphere rise and fall? As ice ages occur in an approximately cyclical pattern, this implies that carbon dioxide levels must somehow vary in a cyclic manner. What can cause this?

Carbon dioxide moves from the atmosphere to Earth through two processes. One is rainfall. The other is the photosynthesis of plants. Carbon dioxide moves from Earth to the atmosphere by respiration of animals, evaporation of water, volcanic eruptions, and fires. Cycles in the density of plant life could, in theory, affect the levels of carbon dioxide in the atmosphere. A high density of plant life absorbs carbon dioxide from the atmosphere, while a low density does not absorb enough to counteract the other processes.

But the biological systems involved are complex. Would an increase in plant life give rise to an increase in plant-eating animals and therefore an increase in respiration? The theory is reasonable, but leaves us with many unanswered questions. Perhaps the biggest question of all is this: Was the drop in the carbon dioxide level during the ice ages the cause of the ice ages? Or was it a symptom—a coincidental result brought about by the real cause?

Volcanoes

Ice core samples show that the air at the time of the ice ages was very dirty and full of debris from volcanic eruptions. Eruptions of volcanoes in the last 200 years and records of worldwide temperatures have shown that volcanoes can affect the weather. The manner in which this happens was discussed in Chapter 3.

Scientists disagree on the question of whether volcanic eruptions can affect the weather to such an extent as to trigger an ice age. Some scientists think so, and point to evidence that high volcanic activity occurred at the start of the ice ages.

But there are many arguments against this theory. Observations of recent volcanic eruptions show that the reflective dust expelled into the stratosphere by volcanoes dissipated in less than fifteen years. These scientists don't believe that even a massive volcanic eruption could cause suspension of particles in the atmosphere for long enough to bring on an ice age.

Another argument against this theory has to do with the character of ice ages, which come on slowly and end suddenly. Volcanic eruptions are, by their nature, sudden, cataclysmic events. Why would sudden, cataclysmic events have effects that are long and gradual?

Another unanswered question involves the cyclical occurrence of ice ages. If volcanic eruptions cause ice ages, why should they happen in cycles? The question is interesting. The dynamics of the plate tectonics that lead to volcanic activity may in fact be cyclical. We don't know enough about it to say definitively that they aren't. But, in all objectivity, we must maintain that the theory, although reasonable, is incomplete, for it leaves many unanswered questions.

Antarctic Glacial Surge

Observations made on the movement of glaciers show some interesting results. Glaciers that move by basal sliding often exhibit a pattern of cyclical expanding and contracting. The pattern can be seen in glaciers that move from mountains into valleys.

As massive amounts of ice move rapidly by basal sliding from high elevations to low elevations, melting becomes more rapid. The glacier retreats. But then, gradually accumulating snows cause the glacier to surge again, and the process repeats.

Can this type of cyclical behavior occur with the huge ice sheets of Antarctica? Normally, these glaciers are frozen to the ground. They cannot move by basal sliding, which requires the presence of a thin layer of meltwater between the glacier and the ground.

But an interesting property of water comes into play here. Unlike many substances, water's liquid state is denser than its solid state. The molecules of liquid water are packed together tighter than the molecules of ice. Ice changes to water under conditions of appropriate temperature and pressure. At normal pressures, ice changes to water at thirty-two degrees Fahrenheit. As pressure is increased, molecules of ice are forced closer together, and the change from ice to liquid water is made easier. Therefore, less heat is required. The melting point of the ice decreases as the pressure increases.

Figure 6-1 is a rough sketch of a "phase diagram" for water. It indicates, for every combination of pressure and temperature, whether the water will exist as a solid, a liquid, or a gas. As the phase diagram shows, as pressure increases, the temperature at which ice changes to

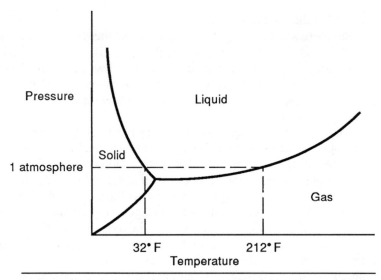

FIGURE 6-1. *A phase diagram for water. Notice that ice can melt at a temperature below thirty-two degrees if sufficient pressure is applied to it.*

water decreases.

This fact has a great deal to do with allowing Antarctic glacial surges. As snow and ice build up atop a glacier, the bottom layers of the glacier feel increasing pressure. Even though the bottom layer of ice is at a temperature well below thirty-two degrees, it can still melt if enough pressure is applied. When it melts, a layer of meltwater is created along the ground, and basal sliding can occur.

The theoretical scenario then goes like this. The ice advances to the ocean and breaks up into icebergs. The total area of the ice on the Earth's surface increases, since it becomes more spread out. Ice reflects such a high percentage of solar radiation striking it, the overall reflectivity (called albedo) of Earth increases significantly. As a result, the Earth cools to form an ice age.

The end of the ice age comes about as the Antarctic glaciers become thinner. The pressure on the bottom layers is decreased. The temperature at which ice melts goes up, and the ice on the bottom layer stops melting. No layer of meltwater is created for the basal sliding. The glaciers stop moving. The icebergs in the ocean gradually melt, and Earth warms up as its albedo is increased again.

The theory is ingenious. It explains the cyclical nature of ice ages in a believable scenario. Unfortunately, the theory predicts that ocean levels should rise at the onset of ice ages, as huge icebergs are thrust into it from the Antarctic glaciers. This does not happen, however, as can be verified from the fossils of corals that can grow only at certain ocean depths, and that verify the sea level as steadily decreasing during all ice age periods.

As original as this theory is, we are forced to reject it on the grounds that it is not consistent with all the available facts.

Meteor Impact

The impact of a meteor onto Earth could throw enough dust into the atmosphere to affect the climate. How a single meteor impact might affect the climate over a long period of time would depend on the material making up the meteor. A meteor made of highly reflective material could block enough of the solar radiation over a long enough period to send the Earth into an ice age.

The process is not that simple, however. The dust blown up by a meteorite made of reflective material would get thrown into the atmosphere. The sun's radiation would be blocked out long enough so that most of the upper atmosphere's water vapor would be turned into ice crystals. The dust would eventually fall to Earth, but the ice crystals, themselves highly reflective, would remain in the upper atmosphere for a long time.

How such an ice age would end is not clear. One proposed scenario involves a second impact of a meteor, this one composed mostly of metal. It would throw darker dust into the atmosphere, which would absorb the solar radiation, heat the surrounding air, and melt the reflective ice crystals.

The theory is lacking, for it fails to explain the cyclical nature of ice ages. Furthermore, it depends on an unlikely chance event for getting out of the ice age. The theory is interesting, and stirs our imagination. But, at best, it is incomplete and unbelievable. At worst, it is not consistent with the available facts.

Solar Radiation

The chemistry of snow during the Little Ice Age and the absence of the northern lights during the same period suggest that some internal mechanism of the sun changed during this period. Fewer sunspots were observed by astronomers of the period.

The same chemical changes of the snow (increase in beryllium-10, which is created by bombardment of electrically charged particles from the sun) occurred during the last major ice age. Could periodic, cyclical changes in the sun's internal workings exist? Could such changes bring on an ice age?

Scientists have identified an approximate eleven-year cycle pertaining to sunspots. The number of sunspots observed on the sun's surface appears to fluctuate in a wavelike pattern, reaching a peak about every eleven years. The observation is intriguing because, even though we can't explain it, it gives the hint of cycles in at least some solar activity.

Scientists estimate that the sun's output would have to decrease by several percent over a few years to lower the Earth's average temperature by the amount necessary to bring on an ice age.

We simply don't understand the sun well enough at this point to say whether this is possible, or whether it can happen in cycles. More research is needed to study the dynamics of the sun and the fluctuations of its internal mechanisms. Until this is done, we must classify this theory as incomplete.

This is unfortunate, because it has a very good chance of being the truth. It is simple, obvious, and consistent with the facts. We just don't have enough information at this point to accept it.

Variations in the Earth's Orbit

Earth's orbit is not constant. Its shape changes over time in a way that is roughly cyclical. "Roughly" is an important word here. Gravitational effects of the other planets and other heavenly bodies cause the orbit of the Earth to fluctuate in a nearly random manner, with exact periodic repetitions virtually impossible.

Nonetheless, mindful of this point, we can still speak of three types of change that the Earth's orbit undergoes, with each change approximating a cycle.

Earth's orbit is not a perfect circle, but an ellipse. The amount of stretch in the Earth's elliptical orbit changes over time, so that the orbit of the Earth varies from a nearly circular shape to a stretched ellipse. When Earth's orbit is stretched the most, the distance from Earth to the sun can differ by about three million miles, depending on the time of year. Currently, the Earth has an elliptical orbit in which it is closer to the sun during our northern hemisphere's winter.

Earth's orbit changes between one that is circular to one that is elliptical in a cyclical pattern. The period of the cycle is about 100,000 years.

The second orbital cycle of the Earth involves its angle of tilt. Earth does not orbit the sun in the same plane as Earth's equator. A picture of this requires three-dimensional thinking. Imagine a piece of plywood cutting the Earth in two right at the equator. Next, imagine a second piece of plywood going through the sun and including all points of the Earth's orbit. The two pieces of plywood make an angle with each other. Currently, this angle is about 23.5 degrees. Earth's orbit is tilted relative to its axis of rotation.

This phenomenon causes seasons. In the northern hemisphere's summer, the northern part of the Earth cuts its orbital plane, and therefore receives the direct rays of the sun. In the northern hemisphere's winter, the southern part of the Earth catches the direct rays.

The angle of tilt varies over time in a cyclical pattern between 21.5 degrees and 24.5 degrees. When the angle of tilt is less, winters are milder and summers are not so warm. As the angle of tilt increases, winters in both hemispheres become more severe, and summers hotter. The angle of tilt fluctuates cyclically over time. The period of the cycle is about 41,000 years.

The third orbital cycle of the Earth has to do with a phenomenon known as *wobble*. Wobble comes about because Earth is not a perfect sphere. As Earth spins, centrifugal force causes it to swell slightly around its equator, so the Earth's equatorial diameter is slightly larger than its polar diameter. As a result, the Earth wobbles like a top as it plummets through space, circling the sun.

Earth's north-south axis, because of the wobble, describes a circle over time. The effect of the wobble is to determine the direction of Earth's tilt at different points of Earth's orbit. It is best understood in conjunction with the elliptic stretch.

The wobble can mean that, when Earth is farthest away from the sun in its elliptical orbit, the northern hemisphere is in its winter. After a while, Earth wobbles so that it reaches the same position during the northern hemisphere's summer.

Suppose Earth is farthest from the sun during the northern hemisphere's winter. Then, in the northern hemisphere, winters will be more severe and summers will be hotter. In the southern hemisphere, at the same time, winters will be milder and summers cooler. Currently, Earth finds itself in the opposite configuration. It is farthest from the sun during the northern hemisphere's summer, and closest during the northern hemisphere's winter. Earth's wobble exhibits a cyclic pattern. Its period is about 22,000 years.

According to the theory, glaciers are formed by the accumulation of ice and snow, which can only occur when summers are not warm enough to melt the snow that falls in winter. Trends that give warm winters and cool summers can lead to the accumulation of ice. The accumulation of ice gives the Earth's surface a higher albedo, so that more of the sun's radiation is reflected back to space. Earth gets colder still, and enters an ice age. The ice age ends when the characteristics of Earth's orbit change back to give warmer summers. Then the ice begins to melt, the albedo goes back down, and Earth warms still more.

The champion of this theory, Milutin Milankovitch, believed that cool summers in the northern hemisphere were the key to the onset of an ice age. The northern hemisphere's seasonal patterns are the most critical with the continents in their current locations. Currently, extended glacial growth can only occur in the northern hemisphere, where the large land masses lie.

Milankovitch computed mathematically the amount of radiation received by various latitudes in the northern hemisphere for different sun-Earth configurations. He concluded that higher latitudes were most affected by the 41,000-year tilt cycle, and lower latitudes were most affected by the 22,000-year wobble.

How does the Milankovitch theory perform? Data of oxygen-16 to oxygen-18 isotopes taken from ice cores and from ocean ooze has enabled us to track the climate of the Earth for millions of years into the past. The climate cycles most closely correlated with Milankovitch's 100,000 year elliptical stretch cycle. The 41,000-year tilt or the 22,000-year wobble seemed to have little or no effect.

The 100,000-year climate cycle was confirmed by data collected from fossils of undersea coral, which showed rhythmic rises and falls of sea level in concordance with the 100,000-year cycle. Fossils of undersea creatures, some of which thrived in warm water and some in cold water, also agreed with the 100,000-year cycle, once non-climatic factors such as water salinity and food supply were accounted for.

It seemed definitively proven that Earth's climate obeyed a 100,000-year cycle, in accordance with the elliptical stretch of Earth's orbit. Core samples taken from all over the world confirm the same 100,000-year cycle, and confirm that the Earth's two hemispheres experienced ice ages at the same times.

Most of the scientific community believed that the mystery of the ice ages was solved, that ice ages were caused by the Earth's 100,000-year cycle of elliptical stretch.

But there were those who disagreed. The main objection was that the 4% variation in solar radiation due to the elliptical stretch was not enough to bring on an ice age. One potent argument pointed to the southern hemisphere, where summers were currently receiving a 3% increase in solar radiation because of the current orientation of the Earth's elliptical orbit. If a 4% warming was supposed to melt the northern hemisphere's ice years ago, why was a 3% warming doing absolutely nothing to the southern hemisphere's ice today? In fact, some scientists today predict an ice age will arrive in a few thousand years based on the wobble cycle.

Others said that Milankovitch had just been lucky. He had picked three horses to bet on in the race, and the third one, his least likely pick, had come in. If you buy enough lottery tickets, you have a better chance to win the prize. Was Milankovitch right or wrong? And how can we find out?

CONCLUSIONS

"I know it's going to rain because my cat is sick. My cat always gets sick right before it rains."

Let's not judge this statement prematurely as a bunch of nonsense.

Maybe it is nonsense. Maybe the speaker of these words tracked the weather for a few weeks, and kept track of hundreds of seemingly unrelated things. Maybe besides the health of his cat, he kept tabs on the number of television commercials between nine and ten at night, the scores of baseball games played that day, the movement of the stock market, or the number of cars going down his street at a certain time.

Out of all the things he looked at, he found one predictor. That predictor was the health of his cat. Was he lucky or not?

If he tries enough things, by random chance, one of them will score a hit. That is basic probability. But can the cat do it again? The cat is sick now, does that mean it will rain?

Let's suppose the cat's health again predicts the rain. Does that mean the owner of the cat was still lucky? Or is there really something to it after all? Maybe low pressure in the atmosphere affects the cat's inner ear, and he gets vertigo. Then, whenever he eats, he throws up. That added explanation makes the theory complete by providing a logical link between the two otherwise unrelated phenomena.

But at the same time, our little example highlights an important principle of science: Correlation does not imply causality. If, each time we see A happen, then we also see B happen, this does not prove that A implies B. Maybe a logical link exists between A and B, and maybe the whole thing is just lucky coincidence. We simply don't know.

Each time we see Earth's orbit reach a certain orientation in the 100,000-year cycle, we also see an ice age. Does that mean one implies the other? The answer is no, not without a firmly established logical link. Others have shown that the variations in solar radiation implied by the cycle are not enough to set off an ice age.

Maybe they are wrong. If they are right, then the needed logical link is not there. Technically, without more data, it is premature to accept the theory of Milankovitch.

And what of the other theories? An abnormal amount of volcanic dust was found in ice cores from the ice ages, low carbon dioxide was also found, and the snow of the time showed a high degree of beryllium-10. There are several other theories (discussed here, the volcano theory, the carbon dioxide anti-greenhouse theory, and the solar radiation theory) that appear consistent with the facts. Why should we choose Milankovitch's theory over these?

The answer is that Milankovitch's theory is the most complete and the most verifiable. This also does not mean it is correct. But we can't explain why volcanic activity should come in cycles, even though we know a lot of it happened at the start of the ice ages. We don't know what could have caused carbon dioxide levels to go down, even though we know they went down during the ice ages. And we don't know what influences the sun to give off less radiation, even though we think that may also have happened during the ice ages. Theories that are incomplete can be neither accepted nor rejected. More research must be done to be able to do that.

The answer is that we simply don't know what causes the ice ages. Maybe it is one of the topics discussed in this chapter, or maybe it is something else completely. A good possibility is that it is a combination of many things. In fact, that is what the evidence suggests. If we see evidence of volcanic eruptions, evidence of lower carbon dioxide levels, and evidence of less radiation from the sun — all when the Earth was in a particular orbital configuration—then maybe this is our solution. Maybe it really is all these things in combination.

There is a particularly intriguing way that we can find out. It is, unfortunately, very expensive, and will probably never happen in any of our lifetimes. But it carries the potential to solve the mystery once and for all.

The idea is to explore the geology of Mars and determine the climate history of Mars. Do major climatic changes on Mars coincide with its orbital fluctuations? If so, then Milankovitch is proven right. Do the climatic changes on Mars correspond with those of Earth? If so, then the answer is in the inner workings of our sun. Is there no relationship whatsoever? If so, then the ice ages are caused by some Earthly phenomenon such as volcanoes or carbon dioxide levels.

Whatever turns out to be the final accepted cause of ice ages, the problem is surely one of the most fascinating in all modern science.

BIBLIOGRAPHY

Books

Bailey, R. *Glacier*. Alexandria, VA: Time-Life Books, 1987.

Barry, R., and Chorley, R. *Atmosphere, Weather, and Climate*. 5th ed. New York: Routledge, 1987.

Chorlton, W. *Ice Ages*. Alexandria, VA: Time-Life Books, 1988.

Cresswell, R. *The Physical Properties of Glaciers and Glaciation*. London: Hulton Ltd., 1958.

Escher, A., and Watt, W. (editors). *Geology of Greenland*. Copenhagen: Geological Survey of Greenland, 1976.

Giancoli, D. *Physics*. 3rd ed. Englewood Cliffs, NJ: Prentice-Hall, 1991.

Hambrey, M., and Harland, W. (editors). *Earth's Pre-Pleistocene Glacial Record*. Cambridge: Cambridge University Press, 1981.

Imbrie, J., and Imbrie, K. *Ice Ages: Solving the Mystery*. Cambridge: Harvard University Press, 1979.

John, B. *The Ice Age: Past and Present*. London: Collins, 1977.

Lehr, P., Burnett, R., and Zim, H. *Weather*. Racine, WI: Western Publishing Company, 1975.

Marcus, R. *The First Book of Glaciers*. New York: Franklin Watts, 1962.

Michael, H., and Ralph, E. (editors). *Dating Techniques for the Archaeologist*. Cambridge: Massachusetts Institute of Technology, 1971.

Pielou, E. *After the Ice Age*. Chicago: University of Chicago Press, 1991.

Schultz, G. *Ice Age Lost*. Garden City, NY: Anchor Press, 1974.

Whipple, A. *Storm*. Alexandria, VA: Time-Life Books, 1984.

Periodicals

Beaty, C. "The Causes of Glaciation." *American Scientist*, July-August 1978.

Broecker, W., et al. "Milankovitch Hypothesis Supported by Precise Dating of Coral Reefs and Deep-Sea Sediments." *Science*, January 19, 1968.

Dansgaard, W., et al. "A New Greenland Deep Ice Core." *Science*, December 24, 1982.

Eddy, J. "The Case of the Missing Sunspots." *Scientific American*, May 1977.

Emiliani, C. "Pleistocene Temperatures." *Journal of Geology*, November 1955.

Evans, J. "The Sun's Influence on the Earth's Atmosphere and Interplanetary Space." *Science*, April 30, 1982.

Fodor, R. "Frozen Earth: Explaining the Ice Ages." *Weatherwise*, June 1982.

Gates, W. "Modeling the Ice Age Climate." *Science*, March 1976.

Maran, S. "The Inconstant Sun." *Natural History*, April 1982.

Neftel, A., et al. "Ice Core Sample Measurements Give Atmospheric CO_2 Content During the Past 40,000 Years." *Nature*, January 21, 1982.

Raisbeck, G., et al. "Cosmogenic Beryllium-10 Concentrations in Antarctic Ice during the Past 30,000 Years." *Nature*, August 27, 1981.

Schultz, P., and Gault, D. "Cosmic Dust and Impact Events." *Geotimes*, June 1982.

Shinn, E. "Time Capsules in the Sea." *Sea Frontiers*, November-December 1981.

Weertman, J. "Milankovitch Solar Radiation Variations and Ice Age Ice Sheet Sizes." *Nature*, May 6, 1976.

Wilson, A. "Origin of Ice Ages: An Ice Shelf Theory for Pleistocene Glaciation." *Nature*, January 11, 1964.

Can Human Beings Travel to the Stars?

You wake up to the sound of light music. You feel groggy at first. It feels like you have been sleeping only a few minutes. You step out of your bunk and walk over to the ship's window. Through the blackness of space, you see the light of a planet. Its continental shapes are far different from Earth's, but certain things are similar. Large patches of blue ocean spread over its surface. Its land areas contain shades of greens and browns. Its clouds are white and fluffy. A printout from the ship's computer tells you that signals have been detected from the planet's surface. The signals display definite patterns and are not at all random. There is a good chance the planet is home to intelligent life.

This is certainly the stuff of science fiction. But could it really happen? Will we ever be able to travel away from our own stellar neighborhood, nestled in a spiral arm of an average-sized galaxy, and reach out toward other planets in other star systems, perhaps meeting with other life forms or other civilizations?

At first, we are tempted to say yes. We can cite the progress made by human beings over the last 100 years, for example, before airplane flight was invented. We can extrapolate into the future along this trend line. The natural conclusion of such reasoning would be to say, "Of course we will do it someday."

But there are many problems, some of them gigantic and some we may never be able to overcome. There are limitations of the human body in terms of life span and response to long periods of near weightlessness. There are psychological problems of being shut off from Earth for long periods of time, and saying goodbye forever to lifelong friends and acquaintances. But the biggest problem of all is the distances involved.

Landing on the moon may have been a giant step for mankind. It was the first time human beings had set foot on another world. The trip to the moon was no small jaunt, either. The moon is about 250,000 miles from Earth. This is about the same distance one would cover by orbiting Earth ten times.

But what is a mere 250,000 miles when compared to the distance to the nearest star? The nearest star to Earth, Proxima Centauri, is about 4.3 light years away. This equals about twenty-five trillion miles. The distance to Proxima Centauri is about 100 million times farther away than the moon. If we think of the distance to the moon as corresponding to a single, rather large step (about 30 to 36 inches), then the distance to Proxima Centauri is equivalent to the distance you would walk by going around the world two and one-half times.

There are only eleven known stars that lie within ten light years of Earth. Some of the brightest stars in the sky are much farther away. Vega (in the constellation Lyra) is about twenty-six light years away. Arcturus (in the constellation Bootes and in a line from the tail of the Great Bear) is thirty-six light years away. Betelgeuse and Rigel, the two brightest stars in Orion, are 500 and 900 light years away respectively. Deneb, the brightest star in Cygnus the Swan, is 1,500 light years away. The center of the Milky Way Galaxy is about 30,000 light years from Earth. The Andromeda Galaxy, one of the closest known galaxies, is over two million light years from Earth. These distances can be described as nothing less than astonishing.

To get a handle on how to solve this distance problem, we first need to understand some physics of motion.

BASIC PHYSICS OF MOTION

A force is a push or a pull on an object. If the force is large enough, it can cause the object to move. Every force can be characterized as having a direction and a magnitude. Forces are thus represented mathematically by *vectors*.

A vector is nothing more than a straight line with an arrow on one end. The length of the straight line represents the magnitude of the force. The direction in which the arrow points indicates the direction in which the force acts.

Some typical forces in the natural world are pushes and pulls exerted by human beings and animals, contact forces of objects, friction, electricity, magnetism, and gravitation.

Newton's Three Laws of Motion

Newton's first law of motion, called the law of inertia, may be phrased like this: "A body in motion or a body at rest tends to remain in that state unless acted upon by some force."

As simple as this first law sounds, it flies in the face of casual observation. If we push a rock through a grassy field, our force will cause the rock to move some distance, but it will quickly come to rest. It will not, on its own, remain in the state of motion, as the first law would seem to suggest.

There is no contradiction, however, when we account for the force of friction. The friction between the rock and the ground acts against the direction of motion of the rock, and so slows down the rock, bringing it to rest.

Newton's second law of motion can be stated as follows: "An object's acceleration is proportional to the net force exerted on it, and inversely proportional to its own mass. The acceleration will be in the same direction as the net force."

Another way of stating this law is through the simple formula $F = ma$. In this formula, "F" represents the net force on the object, "m" represents the object's mass, and "a" represents the acceleration the object experiences. Force and acceleration are represented by vectors.

Gravity is a unique type of force that imparts the same acceleration to all objects it affects. The symbol "g" denotes the acceleration experienced by an object in the gravitational field of Earth. The value of g is about 32 feet per second per second (32 feet/sec^2), or about 9.8 meters/sec^2. The force of gravity acting on an object of mass m is therefore mg, the product of the object's mass and its acceleration.

Figure 7-1 provides an instructive example of an object sliding down an incline. The forces involved are the gravity (mg), the contact force of the incline (C), and the frictional force (f). Notice that the force of gravity can be broken up into two components. One component is parallel to the incline (mg sin A from trigonometry); the other is perpendicular to the incline (mg cos A). The component perpendicular to the incline exactly balances the contact force C, since the object

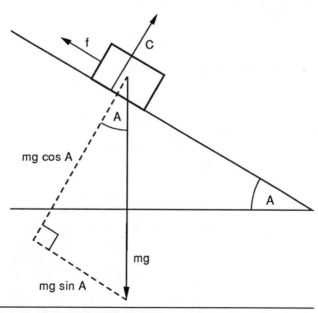

FIGURE 7-1. *Forces involved with an object sliding down an incline include weight (mg), friction (f), and contact forces (c).*

experiences no acceleration in a direction perpendicular to the incline.

If a is the acceleration of the object along the incline, its value can be computed by summing all the forces that act parallel to the incline. These include the object's weight (mg sin A, acting down the incline) and the frictional force (f, acting up the incline). Then, we can write:

ma = mg sin A - f

If we can somehow estimate the magnitude of the frictional force, we can use the above equation to compute the object's acceleration. In fact, frictional forces are well understood, and can be estimated quite accurately from material properties of the substances involved and the weight of the object that is in motion.

Newton's third law of motion may be stated as follows: "For every action there exists an equal and opposite reaction." Newton's third law says that a swimmer pushes water backward in order to go forward, since the force the swimmer exerts on the water is matched by the opposite reactive force of the water on the swimmer, acting to push him forward. Similar logic applies to a rocket, which shoots gases to the rear in order to move forward, or to a rifle, which recoils backward as it propels a bullet forward.

More on Gravity

We have equated the force of gravity to the quantity "mg", where m is an object's mass and g is the acceleration it experiences due to gravity. In fact, this expression is only valid for objects at or near the surface of Earth.

The force of gravity is more complex, and can be stated as follows: Every body exerts a gravitational force on every other body; the magnitude of the force is proportional to the product of the two masses and inversely proportional to the square of the distance between them; the direction of the force lies along the line joining the two bodies.

This can be expressed mathematically as follows:

$F = G M m/r^2$

The symbols have the following meanings:

F = the force of gravity

G = the gravity constant

M and m = the masses of the two objects

r = the distance between the two objects

To use this formula, we must consider all objects to be "point masses." In other words, all objects are modeled as a single point in space with all their mass concentrated at that point. In the case of a person standing on the surface of Earth, we must treat Earth as a point mass, with its mass concentrated at the center. To analyze the gravitational force of a person standing on Earth, we must model the situation as if Earth were a point mass approximately 4,000 miles away (the approximate distance to the Earth's center).

If we use units of kilograms for mass, meters for length, and seconds for time, we have the following:

$G = 6.67 \times 10^{-11}$

M (mass of Earth) $= 5.98 \times 10^{24}$ kilograms

r (radius of Earth) $= 6.38 \times 10^{6}$ meters

The force required to accelerate a mass of one kilogram at a rate of one meter per second per second is called a *newton*. A newton is the basic unit of force in the so-called "k-m-s" system (kilograms, meters, and seconds). The constant G is given the units of (newtons)(meters)^2/ (kgs)^2. These units are assigned to G so that, in the end, everything will cancel out to give the force F in terms of newtons.

Using the numbers for G, M, and r from above, we can calculate the value of GM/r^2 as 9.8 meters/sec.^{2}. This exactly equals the constant "g", which we use as the acceleration due to gravity at the Earth's surface. The force exerted on the object due to gravity is its mass times this acceleration, or mg, or, equivalently, $m(GM/r^2)$, which amounts to exactly the same thing.

The general formula of gravitation, of course, can be used to calculate the gravitational force between any two objects, such as Earth and the sun, Earth and the moon, Earth and Jupiter, etc.

Mass and Weight

The mass of an object is an intrinsic property of the object, and is measured in kilograms. If the object moves from one location to another, its mass does not change.

An object's weight is the magnitude of the force of gravity acting on the object. Your weight on Earth is approximately six times your weight on the moon. By moving from the Earth to the moon, your mass does not change.

Energy and Work

The concept of energy is very important in physics, and it permits us to solve some complex problems with surprising ease and quickness. The tool that allows us to do this is the so-called Energy Conservation Law. There are many ways of stating this law. Here is one way: The change in energy experienced by a system is equal to the amount of work done on the system; if no work is done on the system, then energy is conserved.

To understand this principle, we first need to understand the concepts of energy and work. Energy comes in three basic forms: kinetic energy, potential energy, and heat energy.

Kinetic energy is energy due to an object's motion. There are two types of kinetic energy — translational and rotational. Translational kinetic energy refers to movement in a line or along a path. A marble shooting across the floor and a car speeding down the highway are examples of objects having translational kinetic energy.

Rotational kinetic energy is based on an object's motion due to rotation. The revolving Earth possesses rotational kinetic energy (as well as translational). A spinning top is another example of an object experiencing rotational kinetic energy.

Mathematically, the two types of kinetic energy can be calculated as follows:

K.E.(translation) = $mv^2/2$

(v is the object's velocity and m is its mass)

K.E.(rotation) = $IW^2/2$

(I is called the "moment of inertia" of the body and is a measure of its mass distribution about its axis of rotation; W is called its "angular velocity," which measures the object's rotational speed.)

The calculation of rotational kinetic energy is complex in practice because we need to know the distribution of the object's mass. Translational kinetic energy, on the other hand, is easy to compute.

The total kinetic energy of an object is the sum of its translational and rotational elements.

Potential energy is energy that an object has because of its position. There are many forms of potential energy. The most common is gravitational potential energy. An object suspended at a height has potential energy. If allowed to fall, the potential energy will be converted to kinetic energy. When the object hits the ground, it will possess only kinetic energy and no potential energy.

This example illustrates the idea behind potential energy. Potential energy is energy that can be converted into motion. There are other examples besides gravitational potential energy. An arrow sitting in a stretched bow, or a newly wound mechanical clock, provide such examples.

The gravitational potential energy of an object can be calculated by the expression:

Gravitational P.E. = mgh

(m is the object's mass, g is the acceleration of gravity, and h is its height).

Heat energy is heat that is created due to motion. Whenever an object's motion encounters the resistive force of friction, heat energy will be created. The amount of heat energy created in this way is complicated to quantify mathematically. We will not attempt this, but only refer to heat energy as a general concept.

Work is done by external forces and amounts to adding energy to the system. A human being throwing a football is doing work by giving kinetic energy (both translational and rotational) to the football. Walking up a mountain is work since you gain potential energy by getting to the top. Pushing a heavy box across the floor is work since you are fighting to overcome frictional forces, and are thus creating heat energy.

An example will serve to illustrate the energy conservation law. Suppose an object falls from a height of 1,000 feet. How fast will it be going when it hits the ground?

When the object is suspended at that height, it has no kinetic energy, but it has potential energy equal to mgh. When the object strikes the ground, it has kinetic energy equal to $mv^2/2$ and no potential energy.

The kinetic energy at the bottom may be equated to the potential energy at the top. We obtain the algebraic result that the velocity at the bottom is the square root of 2gh. In our example, with h = 1,000 feet and g = 32 feet/sec^2, we obtain an answer of about 253 feet per second, or approximately 170 miles per hour.

This answer will turn out to be a little high. Our solution method assumed that all the potential energy would be converted into translational kinetic energy. This is not exactly the case. Some friction will be encountered between the object and the air, so some of the original potential energy will become heat energy. The object may also gain rotational kinetic energy. Consideration of these factors leads to extreme complications, since we need to understand how much of the object's rotation is created by work done by the moving air.

Escape from Earth

How fast must a rocket ship go in order to escape Earth's gravity? To answer this question, we must realize that the gravitational force on the rocket ship decreases as the rocket ship moves farther away from Earth. The problem sounds complex, and seems to require integral calculus to take into account the continually changing gravitational force.

But logic surrounding the energy conservation law can be used to avoid these complications. The only trick is that we need to consider Earth as a point mass, that is, a single point with all its mass concentrated at its center. With that in mind, we can use the following energy argument.

The rocket ship starts out with both kinetic energy and potential energy. Its kinetic energy is due to its initial velocity. Its potential energy is due to the fact that it is about 6.38×10^6 meters above Earth's center.

As it moves upward, it loses kinetic energy at exactly the same rate it gains potential energy. This is always true, regardless of the amount of gravitational force exerted on the ship.

Therefore, the only requirement is that the ship's original kinetic energy must be greater than its original potential energy. We can say:

$(1/2)mv^2 = mgR$

(R = radius of Earth)

From this, we can derive the relationship that the initial velocity v must exceed the square root of 2gR in order to escape the gravity of Earth. When we plug in the numbers, we get the so-called "escape velocity" of the Earth as about eleven kilometers per second. This converts to about 6.9 miles per second, or about 25,000 miles per hour.

We can generalize this idea beyond Earth. Thus, if R is the radius of any body in the universe and g is the acceleration due to the body's gravity, we can compute the square root of 2gR to obtain the body's escape velocity. Of course, we may not always know the acceleration due to the gravity of the body, but we can use our previously derived relation that this acceleration equals GM/R^2. Substituting this value for g, we can derive algebraically the escape velocity of any body as the square root of 2GM/R, where G is the gravitational constant, M is the body's mass, and R is its radius.

CONCEPTS IN THE SPECIAL THEORY OF RELATIVITY

Einstein's theory of relativity, along with the theory of quantum mechanics, both coming in the early part of the twentieth century, shook physics at its very roots. Just when it seemed that we humans were very close to a complete understanding of the physical world, we realized how far away we were, and how complicated and non-intuitive the world was. The ideas of relativity and quantum mechanics go against common sense and intuition. Few people would believe in them were it not for the mounting piles of supporting experimental evidence.

Inertial Reference Frames

If you are standing in your backyard while playing catch, you observe Newton's laws at work. You exert a force on a ball to propel it forward, and someone else exerts a force on the ball to stop its motion. Newton's first law certainly holds true. The ball is at rest until somebody exerts a force on it, then the ball moves until someone else exerts another force. Bodies at rest or in motion tend to remain in that state unless acted on by an external force.

You would see the same thing if you were riding on a smooth train at a constant speed. You could play the same game of catch, even though you, your friend, and the ball are all moving at some speed relative to the ground. As long as the speed of the train is constant, you are all right. If the train accelerates, or goes around a bend, you will see the ball behave strangely. If the train is changing speed, the ball will appear to change its speed as you watch it from inside the train. If the train goes around a curve, the ball will appear to curve in its path as you watch it inside the train.

Meanwhile, an observer outside the train will see all the balls thrown move in a straight line with constant speeds. We say that Earth and the straight-moving, non-accelerating train are inertial frames of reference. Newton's first law holds true in such frames of reference. The train that is accelerating or the train that goes around a bend is an example of a non-inertial frame of reference.

In truth, Earth is not an inertial reference frame, since it rotates. Events that take place on a scale that is significant compared to the size of Earth point up this fact. The Coriolis force (discussed in Chapter 6), which causes air to rotate clockwise about a high pressure center in the northern hemisphere, is really not a force at all. It is just the movement of the air through a non-inertial reference frame. The air bends as the Earth spins, just as the ball curves as it is thrown inside a turning train.

The special theory of relativity deals only with inertial reference frames. The general theory of relativity deals with all reference frames, including non-inertial ones.

The Finite Speed of Light

Light travels at about 186,000 miles per second. As we watch events happen around us, it appears they are happening at exactly the moment we observe them. In the world we live in, to a very close approximation, this is true. In the world of the cosmos, the speed of light becomes a limiting factor that is enormously significant.

The speed of light was first noticed by watching Jupiter and its moons through a telescope. As the moons of Jupiter move in back of the planet, they are eclipsed and cannot be seen from Earth. The orbital dynamics of Jupiter's moons are well understood by astronomers. But the behavior of the moons during the eclipses is curious.

When Jupiter is closest to Earth, the eclipses seem to occur earlier than they should. When Jupiter is farthest from Earth, the eclipses seem to occur later than they should. The effect can only be explained by the extra time required for light to travel the added distance when Jupiter is far away. From this data, the speed of light was accurately computed.

Light as an Electromagnetic Wave

In the latter half of the nineteenth century, the physicist James Maxwell predicted the existence of electromagnetic waves. Electromagnetic waves come about because of two interesting facts of physics. The first fact is that a changing electric field produces a magnetic field. The second fact is that a changing magnetic field produces an electric field.

Heated charged particles, such as protons and electrons, which vibrate in simple harmonic motion, create a changing electric field whose intensity varies according to a wavelike pattern. It creates a magnetic field, which is itself changing, and which in turn gives rise to another changing electric field, and so forth.

In terms of the underlying mathematics, the sine and cosine functions that describe simple harmonic motion have an infinite number of derivatives. The derivative of the sine function is the cosine function, and the derivative of the cosine function is a negative sine function. An electric field that oscillates according to a sine function generates a magnetic field that oscillates according to a cosine function (the derivative of the sine). The oscillating magnetic field creates another electric field that describes a negative sine function (the derivative of the cosine). The process continues infinitely, with the result that electromagnetic waves can travel through space.

Maxwell used mathematics and principles of electricity and magnetism to compute the theoretical speed at which the electromagnetic waves would travel. He arrived at the remarkable result of 186,000 miles per second, precisely matching the already known speed of light. It was thus a natural conclusion that light was an electromagnetic wave.

The Speed of Light is the Same for All Observers

If you are on a train going fifty miles per hour and you throw a ball from the rear of the train toward its front at ten miles per hour, an observer outside the train will say that the ball is moving sixty miles per hour. If you throw the ball from the front to the back of the train, the outside observer will say the ball is moving forty miles per hour. But, in all cases, observers on the train will say the ball is moving ten miles per hour. The velocity of the ball depends on the observer's frame of reference.

Amazingly, this is not true of the speed of light. All observers, regardless of their speeds, observe the speed of light to be 186,000 miles per second. The speed of light does not depend on any particular reference frame.

A partial proof of this result comes from observations made on binary stars. (See Figure 7-2.) As one star moves around its companion, its speed relative to Earth changes. We would think that as the star moved away from Earth at a speed v, the light would come from it at a speed c - v (where c is the "normal" speed of light). Similarly, as the star moved toward Earth, we would think that the light from it would speed up, and come with a speed of c + v. But this does not happen. If it did, we would see many double images of the star, as light from the star during different times of its orbit would arrive at the Earth simultaneously. This does not happen, and, in fact, the binary stars orbit their companions exactly as gravitational laws predict, with no necessary adjustment for the changing speed of light.

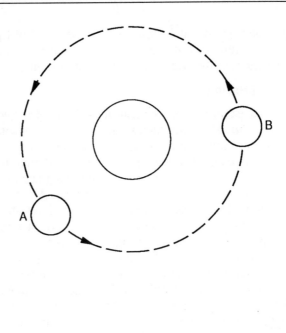

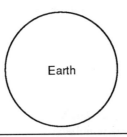

FIGURE 7-2. *A binary star system as seen from Earth. A and B represent different locations of a smaller star orbiting its larger companion. From Earth, we see the star alternately advance toward us and recede from us. The light it emits is not affected by these changes. If it were, we would see a double image of the star as light emitted at different times and at different speeds could arrive at Earth at the same time.*

The famous experiment of Michelson and Morley (described in detail in most physics texts) was a more elaborate attempt to see if the speed of light depended on the reference frame of an observer. The experiment was based on the fact that Earth's speed around the sun is about one ten-thousandth the speed of light. They used beams of light and systems of mirrors to channel the light in various directions. No significant change was observed, whether the light was channeled in the same direction of Earth's traveling, or in the opposite direction.

The only valid conclusion was that the speed of light does not depend on the reference frame of an observer, but is the same for all. The consequences of this are far-reaching and profound.

Time Dilation

To examine the implications of this, Einstein used a technique called *thought experiments.* Since speeds approaching the speed of light were impossible to duplicate in a laboratory, Einstein cooked up fictitious examples that he analyzed to examine his theory. One such example, couched in slightly different language, might go like this.

Imagine you are on the side of a road. A car goes by at a very high speed. Inside the car, a blinking light sends a light beam from the right front window across the car to the left front window. How will the beam of light appear to various observers?

The observer inside the car will see the beam of light move directly across the car in a straight line from window to window. (We are assuming here that the car is an inertial frame of reference; that is, it is moving at a constant speed and not turning. If this were not so, the light beam would be seen to bend in its path through the car.) The observer outside the car will see something different. (See Figure 7-3.) The car will have moved significantly as the light beam travels from

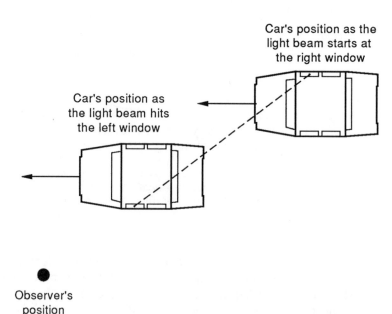

Car's position as the
light beam starts at
the right window

Car's position as
the light beam hits
the left window

Observer's
position

FIGURE 7-3. *A light beam moves across the width of a rapidly moving car. An observer on the outside sees the beam travel a longer distance.*

window to window, and the observer outside the car will see the light follow a diagonal path of increased length.

So the observer outside the car sees the light beam travel a longer distance than does the observer inside the car. But light travels at the same speed for both observers. The only way to explain this is that time runs slower for the observer in the car than for the observer outside. Light travels different distances at the same speed, and therefore must cover those different distances in different amounts of time.

This startling conclusion has been experimentally verified. Atomic clocks flown around the world in high-speed jets confirm the time dilation conclusion. Indeed, the clocks aboard the jets ticked at a slightly slower rate than the clocks on the ground.

The theory says that the moving frame of reference will see time go slower by a factor of $1 - (v^2/c^2)$. Here, v is the velocity of the object and c is the speed of light. The effect is minuscule at normal speeds of everyday life, but enormously significant at speeds that approach the speed of light. A spaceship traveling at 80% of light speed would see time move at 60% the rate on Earth. If it were instead traveling at 99% of light speed, time would move about 5% as fast. Every year spent on such a spaceship would correspond to about twenty years spent on Earth.

Time Dilation is Non-Intuitive

The phenomenon of time dilation does not mesh with common sense or everyday experience. It does not seem to follow logically that just because you are moving fast, time should go by at a slower rate. Yet the theory is accepted as generally correct, and an abundance of experimental evidence supports it. A brief illustrative example points up the reason for the non-intuitive concept.

Suppose you are traveling on a train going fifty miles per hour (about seventy-three feet per second) in a straight line. You throw a ball across the width of the train, perpendicular to the train's motion. This is analogous to the situation depicted in Figure 7-3. Let's say the ball travels ten feet in two seconds. The speed of the ball as you measure it is five feet per second.

An observer on the outside sees the train move 146 feet while the ball is in the air. So the ball moves ten feet in the direction of the train's width and 146 feet in the direction of the train's length. The total distance covered by the ball is then the square root of $146^2 + 10^2$, or about 146.4 feet. The observer on the outside measures the speed of the ball as 73.2 feet per second.

That is no problem. It is all intuitive. The tricky part comes in when we are talking about light. Light has the same velocity for all observers. The person outside the train cannot compute a different velocity for a beam of light than the person inside the train. This is the reason that time dilation is so counter-intuitive. A light beam is not like a ball. If one observer sees a light beam travel ten feet and another observes the same light beam travel 146.4 feet, the two observers can't calculate different speeds for the same light beam. The only way out is to assume that time moves at different rates in and out of the moving train.

Length Contraction and Mass Increase

Length contraction of distances in the direction of travel is a natural outgrowth of time dilation. If a spaceship travels at a given speed for a shortened amount of time, the distance it travels must also decrease proportionately so as to remain consistent with classical physics.

Einstein went on to show that the mass of an object moving at high speed increases and becomes theoretically infinite at velocities approaching the speed of light. He used the known laws of conservation of momentum and the previously determined rules of time dilation and length contraction to arrive at this conclusion. The phenomenon of mass increase for objects approaching the speed of light has been experimentally verified with linear accelerators.

As a final note on this subject, the famous equation relating mass to energy ($E = mc^2$) follows directly from the fact that mass increases with velocity. Only simple algebra is required to derive it.

Is it Possible to Travel Faster than Light?

There are two strong arguments that say it is not possible for an object to travel at precisely the speed of light.

The first argument is from mathematics. If such a thing were possible, the equations of special relativity predict that time would not move, all lengths in the direction of travel would be reduced to zero, and the masses of such objects would become infinite. We therefore argue that such phenomena are clearly not possible.

The second argument is from the physics of electromagnetic waves. If it were possible for someone to travel at the speed of light and ride alongside an electromagnetic wave, the wave would appear stationary to that observer. The observer would see electric and magnetic fields of varying strengths but unchanging in time. In other words, as he looked along the wave, he would see different sizes of electric and magnetic fields, but, as he stood still and focused on one spot, he would see no change. Such a situation is impossible according to electromagnetic theory. It is commonly accepted in scientific circles today that travel at light speed is impossible.

Some have speculated that, although it is impossible to travel at precisely the speed of light, it may be somehow possible to travel faster than light. In the mathematics of special relativity, the quantity

1 - (v^2/c^2) would become negative. Then we must take the square root of this negative number to calculate factors of time dilation, length contraction, and mass increase. The square root of a negative number is called an *imaginary* number. Imaginary numbers have meaning as mathematical constructs, but it is difficult to assign a meaning to them in the physical context of special relativity. Nonetheless, the possibility cannot be ruled out based solely on mathematical considerations.

Physicists have theorized that certain particles called *tachyons* exist, which can travel faster than light, but can never travel at sub-light speeds. The idea is that particles can travel slower than light or faster than light but never at precisely the speed of light.

If tachyons do exist, some think that it is possible to harness them somehow. A highly speculative idea goes like this. A spaceship traveling at sub-light speed is made to undergo an engineering transformation, in which all its particles are converted to tachyons. It then travels at super light speed to its destination, where it undergoes the inverse change to bring itself back to normal.

In spite of long and hard searches for tachyons in laboratories, none have been found or proven to exist. For now the tachyon represents a theoretical concept only. The idea that they can be used to assist humans in traveling faster than the speed of light has to be treated as speculation.

Another idea comes from quantum mechanics. Quantum mechanical theory asserts that energy levels increase or decrease only in discrete jumps, not in gradual increases. (See Chapter 3 for a discussion of this.) Some have speculated that, if a spaceship could accelerate to very close to the speed of light, it could somehow execute a type of quantum jump to a higher energy level, and then start going faster than the speed of light without ever having to travel at exactly light speed in the process.

This idea is imaginative, but it must be said that it is not well understood at this point. Quantum theory is used to describe energy changes at the atomic level. It has been applied successfully to electrons changing orbitals within atoms, but not to entire spaceships changing speeds—a process that would involve the simultaneous cooperation of trillions of such electrons. Although interesting, the idea must be regarded as speculation.

It does indeed seem that there is no way to go faster than the speed of light. If human beings want to go to the Andromeda Galaxy, such a trip would take a minimum of two million years, as measured by clocks on Earth.

PROPULSION SYSTEMS

Every form of propulsion system involves the conversion of energy from one form to another. Early waterwheels took the kinetic energy of running water and converted it, usually to kinetic energy of another kind, to grind grain or forge iron. Windmills accomplished the same sort of things using the kinetic energy of the wind. Modern windmills employ turbines that convert the wind's energy into electrical energy.

A turbine is a very general word that means a device for converting the kinetic energy of a moving fluid into some kind of mechanical energy. Often, the mechanical energy that is the turbine's output is electricity.

A turbine usually consists of rotor blades. A windmill's blades are a typical example. As the fluid strikes the turbine's blades, they are made to rotate and some sort of work can be performed.

The Jet Engine

A jet engine uses a gas turbine, that is, a turbine using gas as the moving fluid to supply energy for the turbine. Figure 7-4 shows a simplified schematic of a jet engine. As the diagram illustrates, the jet engine consists of five basic components: inlet, compressor, combustor, turbine, and exhaust.

The inlet takes in the air from the atmosphere. The compressor squeezes the air down into a much lower volume. This increases the air's pressure, and thus its explosive potential. The pressure of the compressed air is typically about twenty to forty times the pressure of the incoming air. This number is termed the *compression ratio*.

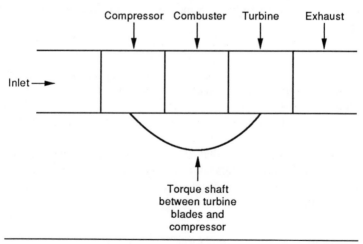

FIGURE 7-4. *Schematic of a jet engine. (See text for description.)*

The combustor mixes the pressurized air with highly flammable liquid fuel. The mixture is burned and then sent through a turbine. The turbine consists of rotor blades and a torque shaft linked to the compressor. The turbine extracts some energy from the heated air and uses it to run the compressor. The air is then expelled through the exhaust. Since it is jettisoned out the rear at high temperature, high pressure, and high velocity, it results in a large amount of force aimed backward. By Newton's third law of motion, the reactive force pushes the jet plane forward.

At very high speeds above mach 3 (three times the speed of sound), a compressor is no longer needed. Sufficient compression is achieved through the "ram" effect, the increase in pressure that results from a jet ramming into the high-speed air. In slightly more technical terms, the velocity of the high-speed air is reduced suddenly as it enters the inlet area. The result is a region of increased pressure. When the speed of the incoming air is above mach 3 or so, the resulting increase of pressure is large enough to eliminate the need for a compressor.

With the need for a compressor gone, there is no more need for a turbine, whose only use is to drive the compressor. The result is a device called a "ramjet," which only requires an inlet, a combustor, and an exhaust. Of course, its limitation is that it won't work at velocities less than mach 3.

The ramjet design can work up to about mach 6. Above mach 6, the incoming air is traveling so fast that even when it slows down in the inlet chamber, it is still moving at supersonic speeds. A very cold fuel, such as liquid hydrogen or methane, is now required for cooling the very hot airplane structure, which is heated by friction with the fast-moving air. The result is a specialized ramjet called a "scramjet" (for supersonic ramjet). The scramjet design, although still existing only in theory, could accelerate a plane all the way to escape velocity (about mach 25).

The combination of jet, ramjet, and scramjet represents a possibility for the not-too-distant future to launch air-breathing vehicles into space. The plane could take off from a runway and accelerate to escape velocity in three stages.

Air-breathers have a distinct advantage over rockets. This is because air-breathers utilize the moving air to drive turbines and obtain most of their energy in this way. Rockets must rely entirely on fuel stored on board. This means that rockets can become very heavy, since such a large amount of fuel must be stored in them. They are also more expensive.

In the vastness of outer space, air-breathers need to be renamed "space-breathers," "interstellar dust breathers," or something along those lines. An interstellar ship that must carry fuel aboard suffers from the serious disadvantage of having to carry such a large volume of material to travel interstellar distances.

Energy from Atoms

Atoms consist of three kinds of elementary particles—protons, electrons, and neutrons—as we discussed in Chapter 3. We know the masses of neutrons, protons, and electrons. We also know the masses of atoms. A curious fact is that the mass of an atom is less than the sum of the masses of its constituents (its protons, electrons, and neutrons). The difference is termed the atom's "binding energy." The terminology reflects the equation $E = mc^2$, which links the ideas of energy and mass. The difference between the atom's mass and the sum of the masses of its constituents measures the strength with which the atom is held together.

We can calculate the binding energy for each atom and then construct the graph of Figure 7-5. The graph shows that the binding energy is maximum for atoms of atomic weight equal to about fifty-six. This element corresponds to iron.

This simple fact has an enormous consequence. It means that atoms lighter than iron can combine to form a heavier element, and, in the process, increase their total binding energy. Similarly, atoms heavier than iron can split apart to form lighter atoms, and, also in the process, increase their total binding energy.

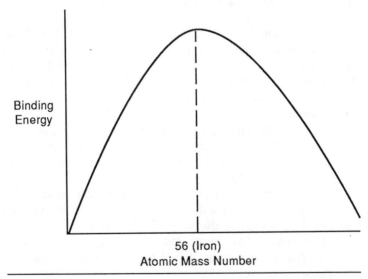

Binding
Energy

56 (Iron)
Atomic Mass Number

FIGURE 7-5. *Sketch of the binding energy for all elements by weight. The curve peaks at atomic mass number fifty-six, corresponding to the element iron. This fact has enormous significance.*

The first case is called fusion. If light atoms can be made to combine to form heavier ones, then there will be an excess of mass that will be converted to energy according to the formula $E = mc^2$. The second case is called fission. If heavy atoms can be split apart to form lighter ones, there will also be an excess of mass that can be converted to energy.

Of course, it is not that easy. In all atoms there are opposing forces trying to rip it apart and at the same time hold it together. The electrical force causes like charges to repel each other. The electrical force is trying to rip apart an atom's nucleus, which consists of many protons of the same positive charge.

Opposing the electrical force is the strong nuclear force. The strong nuclear force acts over very small distances and is an attractive force. Atoms in which the electrical forces of repulsion predominate over the attractive strong nuclear forces are unstable and experience radioactive decay. Their nuclei are unstable and break apart.

In order for fission to occur, the strong nuclear force holding the atoms together must be overcome. Physicists speak of the energy of fission as a figurative hill. At first, energy must be supplied to get "over the hill," that is, to overcome the strong nuclear force. But then, there is a low valley and the eventual energy gain far exceeds the initial investment.

The initial energy to start off the fission process is usually supplied in the laboratory by a neutron that collides with the atom. The kinetic energy of the neutron can sometimes be sufficient to overcome the strong nuclear force of the atom, in which case fission occurs. The target atom splits into two smaller atoms. Usually, one or more neutrons are also expelled as products of the fission.

An example of a fission reaction occurs when a neutron strikes an atom of uranium-235. The outputs of the fission are an atom of niobium-93, an atom of praseodymium-141, and two neutrons. Neutrons, electrons, and protons are conserved. But the mass of the final products is actually less than the mass of the original products because of the greater binding energy of the final products. The result is that some of the mass of the original products is converted to energy.

For fission to occur to a sufficient extent, a type of chain reaction must occur. The chain reaction results from neutrons that are released from a fission. If these neutrons can collide with other atoms, producing more fissions, which in turn release other neutrons, the chain reaction is possible.

The number of fissions occurring in one generation compared to the last is critical. If sufficient neutrons are released so that the number of fissions increases with each step, then a runaway situation can result. An atom bomb is actually a fission device in which the number of fissions in each step increases rapidly and becomes uncontrollable and explosive. The amount of available fissionable material is converted into energy in a very short amount of time, producing a nuclear explosion.

If the number of fissions lessens from generation to generation, the process cannot continue for very long. The result is of no interest to us.

If the number of fissions is about the same for each generation, the result is a manageable process in which energy is created at a stable rate. Such is the idea behind a nuclear fission reactor.

The fission bomb represents one idea for an interstellar propulsion system. A spacecraft would emit fission bombs out of its tail end, which would explode soon after. The shock waves would propel the craft forward. Such a ship could attain speeds of about 3% of light speed, according to estimates by scientists. The method has the disadvantage of rocket technology. All the fuel, in this case the fissionable material, must be carried on the ship as payload.

In the case of nuclear fusion, light atoms are joined together to form a heavier atom. The fusion energy "hill" that must be climbed is the repulsive electrical force of the positively charged protons. For fusion to occur, the force squeezing the atoms together must be sufficient to overcome the repulsive electrical forces.

Fusion is the source of energy of the stars. Gravitational force provides the squeeze necessary to force fusion to occur. Because of the curve's shape in Figure 7-5, no energy can be produced by fusions occurring with iron or heavier elements. The death of a star occurs when there is nothing left for fusion except for these heavier elements.

In the laboratory, fusion is forced by either magnetic forces or laser bombardment. The latter process is commonly termed *inertial confinement*. Unfortunately, so far a successful fusion reactor has not been created. In all attempts, the energy required to force the fusion is not equaled by the energy returned from the fusion. Scientists guess that a working fusion reactor may be possible in the early years of the twenty-first century.

Fusion offers a hope for an interstellar propulsive system. A device using fusion for energy could be "air breathing"; that is, it could use matter found in the interstellar void. It would not have to carry a huge payload of fuel, as would a spaceship operating by fission.

The fuel for such a fusion-powered ship would be the hydrogen atoms found in the interstellar void. The scarcity of the hydrogen atoms places a limitation on this type of power. The front of the spaceship would have to have an enormous cone to trap hydrogen atoms. Some scientists imagine such a cone would have to be over one thousand miles in diameter in order to collect enough hydrogen for fusion to occur at a sufficient rate. Although this represents a significant engineering problem, the so-called "interstellar ramjet" could reach speeds significantly approaching the speed of light, perhaps 50% or even more of light speed. The possibilities are exciting.

Other Propulsive Systems

Many other ideas exist for possible propulsive systems to carry human beings to the stars. Perhaps the most interesting is the "solar sail," a device operating on the same principle as a sailboat.

The force driving the solar sail would be radiation pressure from the sun. The force of this radiation pressure is small, but it increases in proportion to the area of the sail presented to the sun. Counteracting the radiation pressure force is the gravitational force of the sun, which tends to pull the device into the sun. The gravitational force increases in proportion to the device's mass. Therefore, for such a sail to be effective, it must present a very large area to the sun and be very lightweight.

The device could maneuver much as a modern sailboat, using tacking to angle in toward the sun on a return trip. Unfortunately, the necessity for these types of tacking maneuvers places a limitation on such a ship's capabilities. The ship cannot always sail directly to where it wants to go, but must angle in through a series of tacks, thus increasing the distance it must travel by as much as 40%. Another disadvantage of such a device is the limited power it could collect when it is a long distance from the nearest star. On the other hand, the device is simple, is very cheap, and does not require a payload of fuel.

Other types of sailing ships are possible, including those powered by microwaves or lasers fired from Earth or from robotic devices operating in deep space.

Microwaves and lasers have advantages over the solar sail in that they can be aimed in tight beams over millions of miles. Thus, they can effectively propel a ship that is a very long distance away. Some scientists estimate a microwave- or laser-sailing vessel as capable of reaching speeds of up to 20% that of light.

An added complication exists for constructing spacecraft such as the interstellar ramjet or the various types of sails. These ships will not be able to escape Earth's gravity. They will be designed to fly through empty space, not through Earth's atmosphere. Thus, they will have to be constructed in outer space. Construction materials will have to be shipped to the work group, presumably living on an orbiting space station.

OTHER OBSTACLES TO OVERCOME

Unfortunately, the propulsion problem is not the only difficulty associated with interstellar flight. Human beings embarking on this type of adventure will endure great stresses on their biologies and psyches.

The human being was not meant to live in a condition of prolonged weightlessness. Humans on long missions in space have experienced extreme muscle weakness, slowed heartbeats, and calcium loss from bones as a result. Leg muscles begin to atrophy without the stress of having to overcome gravity. The circulatory system also degenerates due to the absence of gravity-induced stress.

The effects of prolonged weightlessness for a period of several years while on an interstellar mission would be catastrophic. The humans would not be able to function at all effectively at the completion of their journey.

Humans on such a long flight would have to do enormous amounts of exercise with devices such as treadmills, bungee cords, and stationary bicycles to maintain any level of fitness. But, as any serious athlete can testify, long periods of regimented exercise stretched over a prolonged time frame can severely challenge one's perseverance.

Other ideas for tackling this problem have centered on ways to induce artificial gravity aboard the spacecraft. An interesting consequence of Einstein's theory of general relativity is that a gravitational field cannot be distinguished from an accelerating frame of reference.

What this really means is that a spacecraft accelerating at a rate of 32 feet per second per second (equal to the acceleration felt by all bodies subjected to Earth's gravity) will feel exactly like Earth. A spacecraft could be designed to accelerate at this rate for exactly one-half of its journey, and then decelerate at the same rate on its second half. The idea is ingenious, but suffers disadvantages associated with loss of flexibility. A propulsive system would have to be carefully chosen and honed, and destinations could not be changed in mid-flight.

A second way to simulate gravity is through a spinning spacecraft. A spacecraft in the shape of a sphere could rotate much as Earth does, creating a centrifugal force, and thus an acceleration against its edges. The idea is quite innovative but suffers from a major disadvantage. Coriolis forces, as experienced by the movements of air masses over Earth, would also be felt quite noticeably on such a device. Since the device would not at all approach the size of Earth, of course, the Coriolis forces would be felt on a smaller scale, and would be quite disconcerting. The simple act of throwing a ball across a room would probably not be immune from some drift associated with the Coriolis forces.

Another design, based on the spinning idea, is a doughnut shape, or "torus." The torus has the advantage of minimizing the Coriolis effect, but still does not seem to tackle all the problems. An uneven weight distribution in the torus, or weight that moves significantly across the torus, could affect rotation and induce a sort of disorienting wobble.

Even bigger problems may not be biological, but psychological and social. How will a large community of people live with each other in conditions of prolonged confinement? What is the optimum type of social or political structure that should be imposed on the travelers? These questions are probably more difficult to answer than those dealing with gravity.

Finally, there is the human life span to consider. Even with ships capable of traveling 20-50% as fast as the speed of light, most trips to interesting stars would take more time than most humans live. Only very short trips could be done in the context of one's life. These "short trips" would still require enormous fractions of one's life. What can be done?

One possibility is for generations to live their whole lives traveling aboard a spacecraft. Eventually, the descendants of those who left Earth would come to the final destination. Many people would live their lives without ever seeing Earth or their destination.

Needless to say, such a solution is asking a lot of people. It is asking a lot of someone to board a spacecraft, knowing that he or she will never see Earth again, and will die aboard the spacecraft. The solution, although possible in theory, seems to suffer from serious sociological problems.

Another solution involves placing the travelers in a state of suspended animation or deep sleep during their trip. The travelers would then wake up at their destination, feeling relaxed and fresh. The idea removes the sociological problems, but requires that all in-flight work be done by computers or robots. Such machines would have to detect operational flaws in the ship and know how to repair them without the benefits of human commands. They would also require the capability to fix problems with each other. The method is certainly possible in the future, assuming continued engineering advances in robotics.

Other solutions are more speculative. One, for example, would require that a drug be developed to retard the aging process. Another solution, dependent on medical technology, states that humans could be kept alive for indefinite time periods by replacing failing body parts with mechanical alternatives. The solution is interesting but a bit bizarre. The original humans, by the end of the trip, would have mechanical arms, legs, hearts, kidneys, and brains. They would actually be better termed *cyborgs*, and would resemble creatures from science fiction. Before closing this section, we should note that even seemingly simple problems of space travel can become complex at velocities approaching the speed of light. A fascinating example is navigation. Stars appear to shift positions to a traveler moving through space at high speeds. Standard navigational ideas based on star sightings would have to be reworked or scrapped altogether. Although this problem is certainly not insurmountable, it illustrates the principle. Even the little things become more difficult when you need to travel at velocities approaching the speed of light.

CONCLUSIONS

There is one overriding question pervading the whole discussion of interstellar travel. Why? Why do it and why bother? Why spend the billions of dollars to make it happen?

From the standpoint of national and international finances, the questions are tough. It is difficult to spend money on such far-out goals in the presence of immediate economic pressures.

The quest for knowledge represents one argument for doing it. It can be argued, on a philosophical level, that knowledge is a worthwhile goal in itself.

A more pragmatic argument involves the ultimate survival of mankind as a species. It is highly questionable whether we can continue to live on Earth for as long as that planet revolves about the sun. Doing so would mean living in the shadow of danger from an array of potential catastrophes. These include toxic pollution, global war, climate change, asteroid impact, or even unforeseen diseases. The long-term survival of humanity can be assured only if the species is dispersed on several worlds, for only then will a portion of it be safe from a single world's disaster.

This chapter is unique from others in the book. The question that it poses is of a unique type. It does not ask about the way things are, or why certain events happen. It asks a question that relates to the abilities of human beings. Can human beings conquer the ultimate challenges posed by the universe in order to survive as a species? This is the question the chapter asks. It is a question about human perseverance and tenacity in attacking problems of incomparable difficulty. It is a question about the strength of the human spirit, and its indomitable will to survive.

BIBLIOGRAPHY

Books

Abell, G., Morrison, D., and Wolff, S. *Exploration of the Universe.* 6th ed. Orlando, FL: Holt, Rinehart, and Winston, 1991.

Adelman, S., and Adelman, B. *Bound for the Stars.* Englewood Cliffs, NJ: Prentice-Hall, 1981.

Cole, D., and Cox, D. *Islands in Space.* Philadelphia: Chilton Books, 1964.

Editors of Time-Life Books. *Outbound*. Alexandria, VA: Time-Life Books, 1989.

——. *Spacefarers*. Alexandria, VA: Time-Life Books, 1989.

——. *Starbound*. Alexandria, VA: Time-Life Books, 1991.

Finney, B., et al. *Interstellar Migration and the Human Experience*. Berkeley: University of California Press, 1985.

Gamow, G. *One Two Three Infinity*. New York: Dover Publications, 1961.

Giancoli, D. *Physics*. 3rd ed. Englewood Cliffs, NJ: Prentice-Hall, 1991.

Herbert, N. *Faster Than Light*. New York: Penguin Books, 1988.

Jastrow, R. *Journey to the Stars*. New York: Bantam Books, 1989.

Lewis, J., and Lewis, R. *Space Resources: Breaking the Bonds of Earth*. New York: Columbia University Press, 1987.

Nicolson, I. *The Road to the Stars*. New York: New American Library, 1978.

Pauling, L. *General Chemistry*. New York: Dover Publications, 1970.

Shadowitz, A. *Special Relativity*. New York: Dover Publications, 1968.

Periodicals

Bilaniuk, O., and Sudarshan, E. "Particles Beyond the Light Barrier." *Physics Today*, May 1969.

Blass, W., and Camp, J. "Society in Orbit." *Space World*, July 1988.

Cassenti, B. "A Comparison of Interstellar Propulsion Methods." *Journal of the British Interplanetary Society*, 1982.

Feinberg, G. "Possibility of Faster-than-Light Particles." *The Physical Review*, July 25, 1967.

Forward, R. "Roundtrip Interstellar Travel Using Laser-Pushed Lightsails." *Journal of Spacecraft and Rockets*, 1985.

Garshnek, V. "Crucial Factor: Human." *Space Policy*, August 1989.

Houtchens, C. "Artificial Gravity." *Final Frontier*, May & June 1989.

Stimets, R., and Sheldon, E. "The Celestial View from a Relativistic

Starship." *Journal of the British Interplanetary Society,* 1981.

Wickelgren, I. "Bone Loss and the Three Bears: A Circulating Secret of Skeletal Stability." *Science News*, December 24 & 31, 1988.

Are We Alone?

Earth is a very special place. It is the only known place in the universe where human beings can wander around minding their own business without needing bulky and complex life-support systems to keep them alive.

Indeed, as we learn more about the universe, we are beginning to appreciate even more the livable conditions on Earth. The universe is full of hostile extremes. The closest planet to Earth, Venus, is a virtual hellhole. Temperatures exceed 1,000 degrees Fahrenheit, and few machines can survive the planet's crushing pressure, which is almost 100 times that of Earth's atmosphere. Other planets and moons in the solar system are either extremely hot or cold, or have atmospheres lethal to humans.

From our cursory knowledge of the small group of bodies in our solar system, it would appear to be an extremely fortuitous event that Earth should contain just the right combination of conditions to make life possible.

On the other hand, there are countless potential planets in the universe, numbering perhaps in the trillions. Even if life is a rare event, given trillions of chances to develop, shouldn't it be fairly common?

Thus goes the debate about life in the universe. The premise that life exists on other worlds can be proven only if we find it. It can never be disproved. Ironically, it is the one scientific mystery of which we have very little grasp.

We approach such a problem using probability theory. We don't know whether life exists, but we try to estimate the likelihood that it does. Even using probability theory, it is difficult to come up with a ballpark estimate for this likelihood. Disagreements between scientists on this are startling. Some say that extraterrestrial life must exist in many places, others say it probably does not exist anywhere.

In this chapter, we pose the following question: What is the probability that extraterrestrial intelligent life exists in the Milky Way Galaxy? We will define "intelligent life" somewhat loosely to mean life capable of communicating through space.

A few words should be said about this question, and its choice of words. Why just the Milky Way Galaxy? Why not the entire universe?

The answer is that other galaxies are millions of light years distant, and, practically speaking, out of communicable range. For example, to send a message to the Andromeda Galaxy, one of our closest neighbors, would take over two million years. Getting a reply would take another two million. Carrying on a conversation would be out of the question, and simply not that interesting.

Why this particular definition of intelligence? Certainly there is intelligent life that has not developed the ability to communicate through space. Dolphins and elephants are examples. The answer is that we are interested in communicating with another life form and sharing information. We are also immensely curious to know if other beings communicate through space or travel through space.

The question, as posed, is specific in what it is looking for. Another way to phrase it might be: What is the likelihood that the Milky Way Galaxy contains other life forms more technologically advanced than us?

If such beings exist, how could we communicate? For about 100 years, radio waves emitted on Earth have been traveling through space in all directions at the speed of light. This is one form of communication. It is not meaningful, however. Another civilization may not be able to decipher these messages.

Meaningful communication must rely on elements of common knowledge existing between us and the other civilization. Such elements include arithmetic, chemistry, and the locations of stars. A lengthy message to another civilization might start out by defining a set of alphabet symbols based on things like prime numbers, atomic numbers of chemical elements, or frequencies of known variable stars (stars that periodically brighten and dim in accordance to a regular pattern). The message would then go on to convey information by the use of this alphabet.

Before proceeding with this chapter, I need to make one very important point. The rest of this chapter estimates the likelihood of advanced life in the galaxy, assuming no intervention from a deity or other supreme being. I am not denying the existence of God. But I am saying that if a supreme being wants to create life, He or She can create the life anywhere, and science can't predict it. This chapter focuses only on the likelihood of advanced life forming naturally, in ways that science can attempt to analyze.

With that in mind, how do we get a handle on this problem? Where can we even start?

THE DRAKE EQUATION—THE BIG PICTURE

The Drake equation provides a rough probabilistic view of how to tackle our problem. One way to write the Drake equation is in the following form:

$$N = R_* f_p n_e f_l f_i L$$

The symbols have the following definitions:

N = the number of planets in the Milky Way Galaxy that are home to intelligent life.

R_* = the rate of star formation in the Milky Way Galaxy, in units of stars per year.

f_p = the fraction of stars that have planets.

n_e = the average number of planets in each such planetary system where the environment allows for the existence of life.

f_l = the fraction of those planets where life actually does develop.

f_i = the fraction of life-developing planets where life evolves to an intelligent level (i.e., where it is capable of interstellar travel or communication).

L = the average number of years such an intelligent civilization can be expected to survive.

The Drake equation gives us the big picture. It tells us how to go about analyzing the problem. If we can estimate the values of the six quantities on the right side of the Drake equation, we can come up with a guess as to the likelihood of intelligent life in the galaxy.

It is within the framework of the Drake equation that we approach this task. We will examine each one of the six terms individually, come up with rough estimates of each, and thus arrive at an estimate for the value of N—the number of planets in the galaxy that are home to intelligent life. As you will see, the analysis of this problem involves a fascinating blend of astronomy, biology, chemistry, probability theory, and even sociology.

RATE OF STAR FORMATION (R_*)

The Milky Way Galaxy is about thirteen to fifteen billion years old. It contains about 100 billion stars. If we assume stars form at an equal rate during the lifetime of the galaxy, simple division can give a rough estimate of about seven stars that form per year.

This estimate is a little low. Stars die out to partially cancel the ones created. So the rate of starbirth is slightly higher. Astronomers generally place this rate at about ten stars per year.

Now here is where it starts to get tricky. Many of the stars that form are totally inappropriate to us; that is, they will never be able to harbor planets that can house intelligent life. To be mathematically correct, we should not count these stars in our calculations. What are the factors we should look at to determine if a star can be counted as favorable?

Stable Lifetime

The sun has been in existence for about five billion years. It probably has another five billion years of stable life ahead of itself. Then it will become very unstable, and thus unsuitable for life. It took intelligent life about five billion years to develop in the sun's neighborhood. We thus require that a star's stable lifetime be about this long or longer to allow for the emergence of intelligent life.

As discussed in Chapter 7, stars operate as fusion furnaces. Stars begin with a collection of hydrogen atoms that are squashed together by the force of gravity. This causes them to fuse into helium. Heat energy is produced by this fusion reaction.

The sun is currently stable. As heat energy is created inside the sun, outward radiation pressure results, which approximately balances the inward gravitational force. Thus, the sun remains approximately the same size as we observe it.

The sun is termed a *main sequence* star because it is currently in the process of fusing hydrogen atoms to make helium. Stars in this stage are stable. At some point, the hydrogen in the sun's interior will all be converted to helium. Fusion will only occur in the sun's outer rim. Then, the sun will swell, as gravitational forces will not have much effect on these far-out rim atoms. The sun's expansion will probably swallow Earth and devour it. When the hydrogen is used up completely, the sun will shrink by gravitational force, until it is again squashed tightly enough to permit the fusion of helium atoms. The sun will then undergo a series of expansions and contractions corresponding to each new fusion step until no more fusion can occur. Then the sun will be dead as an energy source.

The point is this—a star's stable lifetime is that time during which hydrogen fusion is occurring in its interior. The rate at which the core hydrogen is consumed, and thus its stable lifetime, depends upon the size of the star.

Stars are classified by color and surface temperature. They are assigned classes according to the letters of the alphabet O, B, A, F, G, K, M. This sequence of letters can be remembered through the mnemonic phrase "Oh Be A Fine Girl Kiss Me." Table 8-1 summarizes the different star classes along with typical colors, temperatures, and stable lifetimes.

Notice that the color of the star depends on its surface temperature. The frequency of the light emitted goes up with the temperature, as discussed in Chapter 3. This trend can be seen in Table 8-1. The sun is a class G star. Stars hotter than the sun fuse hydrogen at a much faster rate, and thus have a shorter stable lifetime.

From this we conclude that only stars of classes G, K, or M have sufficiently long stable lifetimes to allow for the development of intelligent life on one of their planets.

TABLE 8-1 Star Classifications			
Star Type	**Typical Surface Temperature (Fahren.)**	**Color**	**Typical Stable Lifetime (Yrs.)**
O	50,000	blue-violet	100 million
B	30,000	blue	300 million
A	15,000	blue-white	1 billion
F	12,000	white-yellow	4 billion
G	10,000	yellow	10 billion
K	8,000	orange	30 billion
M	6,000	red	70 billion

Rotation Rate

The commonly held theory about the origin of the solar system is that it began with a cloud of swirling gas and dust. The particles collided and stuck together by gravitational forces, eventually becoming large enough to form planets.

It was fortunate that everything was spinning. As particles orbited the common center at different speeds, because of their varying masses, enough collisions were ensured to allow for the formation of planets. If things were not spinning fast enough, the sun would probably be sitting where it is today surrounded by a disk of gas and dust, instead of by a planetary system.

The original rotation rate of the system is approximately preserved in the rotation rate of the sun. We can also use spectrographic techniques to measure the rotation rates of other stars. Spectral lines of stars are smeared as their rotation speeds increase, while stars with slow rotation rates show sharp, distinct spectral lines. (The spectral effect is discussed in Chapter 3.)

Spin rates are fastest for type O stars and slowest for type M. But it is tricky, and not generally agreed upon, as to what inferences can be made about past planet formation based on current spin rates. Some argue that only stars of types O, B, A, F, and G show sufficient spin rates for planetary systems to have evolved. Others argue that spin rates slow over time, and the formation of planetary systems acts to slow the spin rate dramatically. Thus, the second group argues the opposite: that a currently observed slow spin rate, as seen in stars of class F, G, K, and M, means that planetary systems may have formed at some time in the past. So rotation speed of a star is interesting, but difficult to interpret.

Habitable Zone

Some type of liquid is necessary for sustained life to exist. Scientists are in agreement on this. A human being may breathe in gas, and may have solid bones and teeth. But the chemical and biological processes that keep a human being alive and moving take place in solution. Nutrients are transferred throughout your body by means of a liquid bloodstream. Food is processed in the liver and intestines by chemical reactions that absorb the food into solutions. Solid matter passes through your system only because of the presence of liquid. The same is true for all forms of life we know. Biological processes need some liquid. It is not clear that the liquid has to be water, as it is on Earth. But some liquid has to be there.

Hydrogen is the most abundant element in the universe. The possible liquids that can form using hydrogen and other light elements are water (H_2O), ammonia (NH_3), and methane (CH_4). Not as many solids dissolve in methane as do in water or ammonia, so it would seem that the two most likely liquids on which to base life are water and ammonia. Thus, a star's habitable zone is where either water can exist as a liquid or ammonia can exist as a liquid.

The sun has a habitable zone corresponding to liquid water extending from the orbit of Venus to the orbit of Mars. (Even though liquid water may not exist on Venus and Mars now, they could if conditions on these planets were different.) The sun also has a habitable zone corresponding to liquid ammonia extending from Jupiter through Neptune.

Class O stars have the largest habitable zones. Class M stars have very small habitable zones. It would be rare to find a planet around a Class M star that just happened to lie in the star's habitable zone.

We begin to appreciate the special nature of the sun. It is large enough to permit a significant habitable zone, but small enough to have a long enough stable lifetime for intelligence to develop.

Double Stars

Most stars exist in pairs or triplets, or even clusters. The sun is rare because, as far as we know, it does not have a companion. It is highly unlikely that life could exist in a system dominated by a double, or *binary*, star. The argument to support this is based on the habitable zone of such a system.

For a planet to have life, it must stay in the habitable zone of its star or star system over the full length of time required for the life to evolve. Imagine, for example, that Earth had a highly elliptical orbit.

Suppose that it came as close to the sun as Mercury does, and then went as far away as Jupiter, all in the course of one year. Life would be impossible on it, since it would not remain in the habitable zone of either liquid water or liquid ammonia. No important liquid would be able to remain a liquid for very long.

A planet in a binary star system would suffer the same type of problem. For example, imagine that Jupiter was a burning star instead of just a large planet. Then Earth would experience wild variations in the heat it received, depending on whether Jupiter was on our side of the sun or on the opposite side. Water would probably boil away during much of the year, and we would not be able to live.

The Importance of Heavy Elements

Were it not for the fusion taking place within stars, there would be no element in the universe except for hydrogen. A star forms heavier elements in the fusion process. When the star dies, the store of heavier elements is dispersed throughout space.

The oldest stars of the Milky Way Galaxy, those near the Galactic hub, are the so-called *first generation* stars. They were formed from hydrogen. Any planets surrounding them were formed out of hydrogen. Life on any such planets is impossible, since either oxygen (to form liquid water) or nitrogen (to form liquid ammonia) is needed.

It is also generally agreed that carbon is necessary to support life. This is because of the rich amount of chemistry carbon-based compounds allow for.

Since stars only fuse elements up to iron, the reader may now wonder how elements heavier than iron can exist in the universe. The answer lies in a phenomenon called a *supernova*. A supernova is a violent explosion experienced by some stars during death. But the forces that occur during the supernova explosions are so great that they immediately fuse many atoms together, creating the heavier elements, including those that are heavier than iron.

These heavy metals are not important to life itself, but they are important to the development of intelligent life. In our own human experience, a little reflection will tell us the importance of heavy metals. Without metals such as tin, lead, zinc, nickel, and copper (all heavier than iron), we would not have been able to develop a technological society.

As discussed in Chapter 3, much of Earth's internal heat comes from radioactivity in Earth's crust. This radioactivity would not be possible without the presence of heavy elements.

There is one inescapable conclusion: intelligence on Earth only came about because of previous supernova explosions in other parts of the galaxy. The primordial cloud that formed the solar system must have contained the heavier elements in the remnants of earlier super nova explosions. Otherwise, life on Earth could not have advanced to its present stage of intelligence.

Summary

If we want to look for stars in our galaxy that house intelligent life, we should look to the spiral arms, for those new stars formed there will contain the products of earlier supernova explosions. We should look for stars like our sun, large enough to support a significant habitable zone, but small enough to permit a long stable lifetime. We should rule out stars that are members of binary systems or larger close groups.

Mindful of these considerations, what value should we pick for R_*? Earlier, we estimated ten stars per year that were being created throughout the galaxy. But if we only count those stars that have some chance of producing intelligent life, then the value we must use is substantially less. I will use a value for R_* of 0.2.

THE FRACTION OF STARS THAT HAVE PLANETS

In our model, the fraction of stars that have planets is the quantity f_p. Estimating f_p can only be done by rough guesswork, since we have not yet found any star besides our own that has a planetary system. On the other hand, planets are very difficult to detect, and the fact they haven't been found does not prove their nonexistence.

As we discussed earlier, the rotational speed of the system at the time of its creation probably affects the likelihood of planetary formation. Unfortunately, this information is very difficult to infer from current observed rotation speeds of stars, and no firm estimate can be made based on it.

Planets around other stars can be sought out through observational techniques. This is extremely difficult. One such technique is to watch a star for a type of effect known as a *gravitational wobble*. The wobble can occur if a star is surrounded by large planets that exert a gravitational pull on the star.

Another way to think about this is that any pair of bodies exerts a gravitational force on each other. The moon exerts a gravitational force on Earth, just as Earth exerts a force on the moon. Because of this, Earth and the moon actually both revolve about a point that is the center of mass of the Earth-moon system, called the *barycenter*. Earth is much heavier than the moon, of course, so the barycenter is much closer to the Earth's center than to the moon's center. In fact, the barycenter of the Earth-moon system actually lies in the interior of the Earth, but is offset considerably from the Earth's center. Earth, as it circles this barycenter, describes a type of unbalanced wobble.

In theory, the wobble effect can be observed by watching the motions of stars. If a wobble is seen, it is possibly caused by a large planet or a planetary system. Of course, there are other possible causes of a wobble. One is the presence of an object called a *brown dwarf*. A brown dwarf is a star that never became quite large enough for fusion to occur. Brown dwarfs certainly exist, but there is wide disagreement among scientists as to how populous they are.

A second way to try to detect planets is to look for changes in a star's velocity as it circles the barycenter. Velocity changes can be easier to see than position changes by using the phenomenon known as *Doppler shift*. As an object emits waves, the wavelength of the waves will be shortened as the object moves toward us, and lengthened as the object moves away. It is by this Doppler shift that an approaching train sounds far different than one that is moving away.

We can imagine the train's sound waves as consisting of a series of peaks and troughs. Let's say that the train emits one peak per second (for simplicity, not accuracy). As the train approaches us, it moves a significant distance during the one second between the peaks. We will see the next peak "squashed" a little up against the one before it, since the train is moving. As the train recedes away from us, the peaks will be stretched and appear farther apart. Our observation is a change in the wavelength (and also the frequency).

When we observe a star, we can see a slight color change corresponding to a change in the frequency of the waves emitted if the star is moving toward us or away from us. In this way, we can deduce a star's velocity or a change in velocity. If we watch a star carefully over time, we can detect it as it moves around a barycenter. We can then conclude with caution that either a planetary system exists or maybe another object such as a brown dwarf is present.

Another method to detect planets is to observe the infrared radiation that a planet gives off. Unfortunately, stars also give off infrared radiation, so it is difficult to sift out what is coming from the planet and what is coming from the star. But the method does hold promise given future technological advances in instrumentation.

Sometimes we can observe disks of gaseous material surrounding a young star, and guess that planets will eventually form in those areas. Again, it is not easy to observe these disks, although some have been found.

Finally, if we are extremely lucky, we may be able to spot an eclipse. One of the star's planets may pass between the star and Earth, thus eclipsing the star from our view for a short time. Such an event would be extremely lucky, and would require that the star's planetary plane is aligned with Earth. Such an eclipse has never been observed.

Some astronomers maintain that all stars will form planetary systems, and assign f_p the value 1. I believe this to be wildly optimistic, especially in view of the fact that no extrasolar planetary system has yet been discovered. Considering that we have yet to find any definitive proof of another star containing a planetary system, any estimate of f_p should respect this total absence of evidence. Any estimate should be conservative to reflect the poor state of our knowledge on this subject. I will use a value of f_p equal to 0.5.

NUMBER OF PLANETS POSSESSING A LIVABLE ENVIRONMENT

This is the quantity n_e in the Drake equation. Considering only those stars that have a chance to yield intelligent life (those that we used in our calculation of R_*), and considering only the stars out of that group that contain planets, how many planets on the average will each of those stars have that are in the habitable zone? That is a long question, but it expresses what we are after.

Again, the answer to this involves guesswork. We have no other experience outside our own solar system. In our own solar system, we can argue that Venus, Earth, and Mars lie within the habitable zone for water, and Jupiter, Saturn, Uranus, and Neptune lie in the habitable zone for ammonia. If we allow for the possibility of intelligent life to develop on satellites, then we can have in total close to forty worlds within the sun's habitable zone.

Of course we can argue that there is a little more needed than just being located in a habitable zone. The strong gravity of the large planets would make intelligent life difficult on them. Tidal forces exerted by the large planets on their moons are so great that it would make intelligent life very unlikely on those hostile worlds.

Finally, the outer worlds of the solar system are much less metal-rich than the inner worlds. This is because, as the solar system formed, heavier, denser metals were drawn by gravity toward the inner portions. So, considering the requirements of metal-richness, reasonable gravity, and absence from abnormal tidal forces, we are probably down to three worlds in the solar system where life could have started. We are looking at Venus, Earth, and Mars. I will use a value for n_e of 3.

THE LIKELIHOOD OF LIFE DEVELOPING

On a planet with an environment that is suitable for life, what is the probability of life developing? This is the quantity f_l in the Drake equation. This question is too complex to answer in just a few pages. The best I can attempt to do is to discuss various theories of how life could have begun and attempt to give some assessment of the process's repeatability. The estimate that I will have to offer of f_l will of course be very rough.

How did life begin on Earth? That question in itself is a profound mystery. We must attempt to answer it, however, for the answer to that question provides us with our only clue toward determining the likelihood of extraterrestrial life. In the broad sense, there are two theories of how life began on Earth. The first, called the *panspermia* theory, is that life developed elsewhere in the galaxy, and the seeds of life were transported to Earth by means of comets, meteorites, or interstellar dust.

The panspermia theory turns on the believability of microorganisms being able to survive the intense cold of interstellar space, as well as the bombardment of cosmic rays and potentially lethal X-rays to which they would be exposed without the protection of an atmosphere. Partial support for the panspermia theory comes from our own lunar experience. Microscopic bacteria, transported by us to the moon by accident, were later found still alive and thriving three years later. They had managed to live and reproduce in the harsh environment of the moon.

Ironically, however, this observation also provides fuel for the critics of panspermia. For we know that the rest of the moon is barren of life, including bacteria. Now, according to panspermia, if the seeds of life drifted through space and landed on Earth, some of them must have also landed on the moon. But they survived on Earth and not on the moon. Yet our experience says that perhaps primitive life can survive on the moon. This puts panspermia in a difficult spot.

Another notion is that of *directed panspermia*. According to this highly speculative idea, advanced beings from another world visited Earth a long time ago, bringing with them the seeds of life. Perhaps they deposited them intentionally or left them behind as trash from their picnic. Although directed panspermia cannot be disproved, it must be regarded as highly speculative and dubious.

The final notion is that life developed spontaneously here on Earth. Scientists differ on the likelihood of this occurring repeatedly on other planets. At one extreme, some say that the probability of spontaneously generated life anywhere is so small that, if it did occur on Earth, then Earth probably contains the only life in the universe. Others, at another extreme, say that life is probably very common throughout the galaxy. They point to the fact that Earth is about 4.5 billion years old, and the first signs of primitive life are from about 3.8 billion years ago. Life may have even been around earlier than that, but the point is that life developed on Earth fairly quickly, by astronomical time standards. Thus, the spontaneous generation of life is not something unusual or difficult, according to this second school of thought.

Whether it is something that can easily happen, the chemistry of life is quite complex, as we will now illustrate. Some background knowledge is necessary.

How Atoms Combine

Atoms consist of a nucleus of protons and neutrons, surrounded by a cloud of orbiting electrons. The electrons orbit at several different distances from the nucleus. The electron orbitals represent energy states of the atom. An electron can move from one orbital to another if the atom gains or loses energy. Orbitals that are farthest from the nucleus correspond to high energy states. Thus goes the theory of quantum chemistry. The theory is quite powerful. It explains the series of spectral lines that correspond to each known element. (See Chapter 3.) Atoms absorb energy, allowing their electrons to jump to higher orbitals. As electrons fall to lower orbitals, the atom radiates energy. But the difference of the orbitals is always the same. That is the rule of quantum chemistry. Only certain orbitals, or energy states, have the right to exist. For each type of atom, there exists an absorption spectrum and an emission spectrum. These spectra correspond to the possible energy changes that can result from the atom either absorbing or emitting energy. From the basic premise of quantum chemistry, that only certain energy states are allowed, it is obvious that the absorption and emission spectra of an atom are identical.

Electron orbitals are grouped according to their distances from the nucleus. Groups of orbitals with equal radii are termed *shells*. For example, in the chemical element sodium (symbol Na), each atom has three shells of electrons. In sodium's normal configuration, there are two atoms in its inner shell, eight in its middle shell, and one in its outer shell, making eleven in all.

In the element chlorine (symbol Cl), each atom also has three shells of electrons. In the normal configuration, chlorine has two electrons in its inner shell, eight in its middle shell, and seven in its outer shell, making seventeen in all. The reader may note that the atomic numbers of sodium and chlorine are respectively eleven and seventeen, corresponding to the number of electrons found in each.

Sodium and chlorine like to combine to create sodium chloride, which is the chemical name of normal table salt. Why? The key idea is that the third shell of an atom has the capacity to hold at most eight electrons. If chlorine had just one more electron, it could fill up its third shell. If sodium had just one less electron, it could get rid of its third shell and have a full second shell.

For both atoms, this is a good deal. A particularly stable configuration for an atom is when its outer shell is full. Sodium is happy to give up an electron, and chlorine is happy to receive one. As a result, sodium loses an electron, thus acquiring a positive charge, and chlorine gets an electron, thus acquiring a negative charge. There is now an electrical attraction between the two unlike charges, and the atoms stick together. This basic type of chemical bond is called the electrostatic bond.

Another, stronger type of bond comes about when electrons are shared between two atoms. It is termed the *covalent* bond, and is based on exactly the same principle as described above. Atoms try to end up somehow with a full outer shell of electrons. The difference between covalent bonding and electrostatic bonding is that in covalent bonding, electrons are shared between multiple atoms, not just passed from one to another.

An example is carbon dioxide (CO_2). Carbon has two atoms in its inner shell and four in its outer shell. Oxygen has two and six, respectively. Carbon needs four more electrons to make a complete outer shell. Each of the oxygen atoms needs two. The trick is accomplished by the covalent bond, as shown in Figure 8-1. Each oxygen atom shares two of its electrons with the carbon atom. The carbon atom in turn shares two electrons with each of the oxygen atoms.

The vast majority of chemical bonding occurs by the above two methods.

Importance of Carbon

The *valence* of an atom is the number of electrons needed to fill an outer shell. The valence can be negative to represent an atom that needs to lose electrons to end up with a full outer shell. For example, the valence of oxygen is +2, the valence of sodium is -1, and the valence of chlorine is +1. The valence of carbon can be expressed as either +4 or -4. Carbon is therefore a bit special, although certainly not unique, in that it can gain or lose electrons with equal ease, depending on the situation. Carbon is very flexible as an

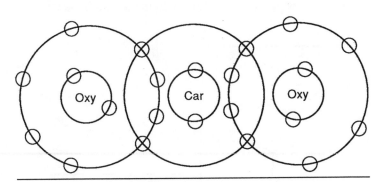

FIGURE 8-1. *The covalent bond as illustrated for a carbon dioxide molecule.*

element to combine with others.

In fact, it is because of this that carbon is the basis of life. It can combine with so many things that the molecules it forms can become very large and complex to fit the needs of advanced life forms.

Of course, carbon is not the only chemical element with a valence of +4 or -4. Silicon (symbol Si), with an atomic number of fourteen, also has four electrons in its outer shell. Silicon's outer shell has a capacity for eight electrons, so it too has a valence of either +4 or -4. Scientists have often speculated that silicon could also be the basis for life because of this same property that it shares with carbon.

There is an important difference, however. A carbon atom only has two electron shells, while a silicon atom has three. Thus, the electrons of silicon's outer shell are farther away from the nucleus than the electrons of carbon's outer shell. The difference is significant because of the electrical attractive force between an atom's positively charged nucleus and its negatively charged electrons. The strength of this attractive force decreases with the square of distance (exactly as the gravitational force). Thus, the forces that pull on the outer shell of carbon are stronger than those that pull on the outer shell of silicon. The bonds that carbon forms are therefore stronger and more stable than the bonds of silicon. Thus, carbon is the preferred base chemical element from which the complex molecules of life can be built.

Amino Acids and Proteins

Amino acids, often called the building blocks of life, are nothing more than simple compounds involving carbon. One of the simpler amino acids is cysteine. It is shown schematically in Figure 8-2. Here, each C stands for a carbon atom, O for oxygen, H for hydrogen, and S for sulfur. As you can see, there is not much to cysteine. In all, it takes just fourteen atoms to make one molecule of it.

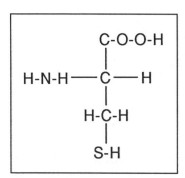

FIGURE 8-2. *A schematic of the simple amino acid cysteine.*

Scientists have theorized that amino acids formed early in Earth's history. They have suggested that energy from lightning or from volcanic heat was sufficient to energize chemical reactions that formed amino acids from simple chemical elements. The idea has since gotten credibility from laboratory experiments. American Stanley Miller simulated conditions of Earth's early atmosphere by enclosing a mixture of methane, ammonia, water vapor, and hydrogen in an airtight terrarium. When electrical discharges, simulating lightning, were passed through the terrarium, amino acids were formed.

Others have cited evidence that comets and meteorites have transported amino acids to Earth. Indeed, it appears that the easiest of all of life's problems is explaining the arrival of amino acids. They are simple substances. At least two methods have been suggested that explain their presence on Earth.

There are a large number of amino acids possible, but only twenty are actually used as building blocks for life on Earth. The twenty amino acids string themselves together to form *peptide* chains, which can be visualized as long writhing snakes. Peptide chains tangle together into knots like many pieces of twisted thread, thus forming proteins. Proteins are extremely complicated substances at the atomic level. A single protein consists of thousands of atoms.

Proteins are the substances that make advanced life possible. Antibodies that protect us from disease are really nothing more than proteins. Proteins called *collagens* provide structural support for our bones. The protein called hemoglobin performs the very necessary function of transporting oxygen throughout our bodies. And with hemoglobin comes the first evidence of the necessity human beings have for supernova explosions. For each molecule of hemoglobin contains an atom of iron, most likely produced and transported through space by earlier supernovas. There are millions of proteins, each serving a different purpose in the role of life support.

We can appreciate the complexity of proteins and the large diversity of them by contemplating their makeup. The proteins found in living organisms are composed of the twenty different amino acids, repeated many times and strung together in a particular set of strands. If each of the amino acids were a letter of the alphabet, a peptide chain would represent a word, and a protein a sentence. The number of amino acids used for life on Earth (20) is conveniently close to the number of letters in our alphabet (26), so that the comparison has some validity.

You can think of the number of possible proteins as the number of possible sentences in the English language. The number is so vast, we cannot even guess at what it is. Moreover, most such sentences are "nonsense" sentences. They make no sense. It is an extremely rare accident that we can string together random letters into random groupings, and somehow come up with a real sentence having a subject and a verb. It is even more of an accident to find a sentence that not only makes sense but is intelligent; that is, conveys valuable information to help us somehow. It is accidents of this sort of magnitude that brought the right amino acids together in the right peptide chains, and ordered the peptide chains in a way that made sense, and finally made it possible for our advanced form of life to exist.

DNA and the Replication of Proteins

A major requirement for life is that it be able to reproduce. A newly created life form must be able to form the same set of proteins as its parent or parents. The manner through which this is accomplished is through the intricate molecule called deoxyribonucleic acid, or DNA for short.

DNA contains a set of plans for reconstructing an organism's set of proteins. DNA is an extremely intricate construct. It is a wondrous phenomenon of nature that a construction-type blueprint can be created by using purely biological and chemical processes.

DNA is often referred to as the "double helix," an expression that describes its unique shape. A helix is another word for a spiral, a type of twisting staircase. Imagine two of the helix shapes next to each other, intertwined over a long distance, and you have the shape of the DNA molecule.

The DNA molecule exists in three dimensions. Figure 8-3 illustrates how the DNA molecule is composed. It cannot represent three dimensions, however, because of the flatness of paper. So the figure shows DNA as if it were squashed flat against the page.

The DNA molecule consists of three components: phosphates (chemical compounds involving the element phosphorous), sugars (carbon-based compounds), and bases (also carbon-based com-

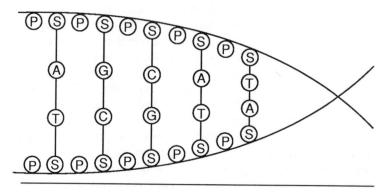

FIGURE 8-3. *A schematic of a portion of DNA strand.*

pounds). The critical parts of DNA are the bases. The ordering of the bases specifies the genetic code, as will be discussed below. The sugars bind the bases to the DNA strands. Finally, the phosphates hold the sugars in line on the DNA strand.

There are four possible bases in DNA, called thymine (T), adenine (A), guanine (G), and cytosine (C). The bases are really just simple compounds involving carbon. The bases combine with each other, reaching across the strands as shown in the figure and linking the two strands of DNA together. The critical point here is that base T always bonds to A, and G always bonds to C. There can be no exceptions to this rule.

The particular sequence of bases on one of the DNA strands uniquely determines the sequence of bases on the other strand because of the limitations on the way the bases can link up. Because of this, the particular sequence of bases on one DNA strand can uniquely identify the DNA molecule. Thus, a DNA molecule is often referred to by its "spelling sequence." For example, CACGATG . . . might represent the beginnings of such a sequence, with each letter identifying one of the four bases in a sequence.

The spelling sequence of DNA is grouped by threes. Each group of three such "letters" of the spelling sequence is called a "codon." Each letter can have four different values—A, T, C, or G. Thus, each codon can take on 4^3, or 64, different possible values. This is actually a bit of overkill, since each codon represents an amino acid, and only twenty amino acids are ever used in the construction of proteins. Nonetheless, three is the minimum number of letters needed, since two letters per codon would only allow for sixteen unique amino acids to be specified, which is insufficient.

Codons are in turn grouped to specify proteins. The codons that specify a specific protein are set aside by a special "starter" codon to begin the sequence, and a special "terminator" codon to end the sequence. In this way, the group of amino acids can be specified one after another in a string.

The process that interprets DNA and uses it to create proteins is nothing less than remarkable. I will attempt to describe it here in a grossly simplistic form.

The process involves, besides the DNA itself, three other chemical constructs:

(1) *Messenger RNA* (RNA is short for "ribonucleic acid"). Messenger RNA is assembled from the information encoded in DNA. Messenger RNA contains information needed to construct one specific protein molecule. It is needed as a carrier of the DNA information in a type of "scratch" area. The messenger RNA will be gradually disassembled during the protein-building process. This is why DNA itself cannot be used. The information that it contains must be copied

into individual messenger RNA's, one messenger RNA for each protein to be assembled.

(2) *Transfer RNA.* Transfer RNA is used specifically to transfer amino acids from the "warehouse" to the "assembly area." The warehouse here represents the body of a cell, and the assembly area represents our third chemical construct, called a "ribosome."

(3) *Ribosomes.* Ribosomes are, as mentioned above, a type of assembly area for building proteins. They read codon sequences from messenger RNA, put in requests for amino acids, and attach the amino acids to a growing chain to form a protein.

The process can be summarized as having the following steps:

(1) A strand of messenger RNA is assembled. Codons in the DNA are scanned for a starter codon, which signals the beginning of an amino acid sequence for a particular protein. When the starter codon is found, the codons following it are encoded into the messenger RNA. The encoding process stops when a terminator codon is detected.

(2) The messenger RNA, which now contains all the necessary information to form a specific protein, enters the interior of a ribosome.

(3) The ribosome analyzes a codon to determine which amino acid is being specified.

(4) The ribosome puts in a request for delivery of that amino acid. The request is answered by transfer RNA, which brings the requested amino acid to the ribosome.

(5) The ribosome takes the amino acid from the transfer RNA and attaches it to a growing peptide chain, which will eventually go into forming a protein.

(6) The ribosome continues with steps (3) through (5) until the information contained in the messenger RNA is exhausted. The formation of the protein has been completed.

(7) A new step (1) is performed for the next sequence of protein-specifying codons. The whole process continues through the entire DNA strand.

In reading the above process description, one is struck by the fact that it sounds very intelligent, akin to a series of steps programmed on a computer. In fact, the comparison is not too far off. Messenger RNA can be thought of as magnetic tape. A ribosome can be thought of as a combination magnetic tape reader and memory manager. Transfer RNA corresponds to readable memory. The peptide chains correspond to writable memory. So the ribosome reads the instructions on the tape (messenger RNA), then performs a type of memory move from a readable area (transfer RNA) to a writable area (the growing peptide chain).

Of course, the remarkable thing about this apparently intelligent process is that it is not intelligent at all. In fact, it bears no resemblance at all to a computer system. It is chemistry. The forces that drive the process are chemical reactions taking place in solutions, nothing more. The fact that a set of purely chemical processes can lend the appearance of such a high level of intelligence is one of the true wonders of the scientific world.

Genes, Chromosomes, and Cells

A grouping of codons within a DNA molecule, which is sufficient to specify one protein, is called a gene. A collection of codons specifying a large number of proteins makes up a chromosome.

There are thus two ways to look at a chromosome. One is as a collection of genes. A second way to look at it is as a long strand of DNA. Each chromosome consists of approximately 100 million "nucleotides." A nucleotide is just a fancy word that means a letter representing one of the bases (i.e., A, C, T, or G) in DNA. The number of genes contained in a chromosome varies, depending on the lengths of the proteins encoded within the chromosome's DNA.

Most proteins contain over 100 amino acids. If we use 100 as an approximate working number, and remember that three nucleotides are required to encode each amino acid, we have about 300 nucleotides per protein. A chromosome containing 100 million nucleotides would then have the capacity to hold about 300,000 genes (100 million divided by 300).

Every human being has forty-six chromosomes, or about fifteen million genes. Each gene can have countless configurations of amino acids, so the total number of unique human beings is an astonishingly huge number.

Chromosomes are housed within cells, which are the basic units of life. In advanced life, cells become specialized. In humans, there are liver cells, brain cells, blood cells, and countless others that assume very specialized functions. Each cell receives the full amount of chromosomes, but not all of the genetic information is used by each cell. The cells in the liver only need to know the information concerning liver functions, not about hair color, for example. The manufacture of proteins necessary for a liver cell, therefore, only involves those proteins needed to perform the functions of the liver.

In asexual reproduction, offspring receive exact copies of the parent's chromosomes. Thus, there is no variability of such life forms from generation to generation except for relatively rare instances of copying errors.

In sexual reproduction, we can think of the offspring receiving half of his chromosomes from each of the two parents. It is a useful way to think about it, but it is actually a bit of a simplification. What actually happens is that the chromosomes in each parent cell break at random points in their middles. When the cells of the parents unite, the broken chromosomes recombine with those of the opposite parent. A type of random shuffling of the genes is thus created. The recombination process is tricky, since the offspring must eventually end up with a full set of forty-six chromosomes, with each chromosome containing a full set of genes. The exact way the broken chromosomes are recombined is not completely understood.

An interesting result of the broken chromosome phenomenon is that genes that are physically located close to each other on the same chromosome are likely to be passed as a pair to the offspring. This is because the chromosomal break is more likely to separate genes that are far apart than those that are close together. The fact has interesting implications to genetic research and the hunt for certain genes that cause birth defects and other genetic disorders.

The number of chromosomes endowed to each living thing is a measure of the being's complexity level. Most living things have far fewer chromosomes than forty-six, the number allocated to humans.

Summary

The last few pages have attempted to show the complexity of the processes needed to sustain life. The central question of this section is to estimate the probability f_l that life can form (without divine intervention) on a planet that is home to a suitable environment.

We may be no closer to answering this question than we were at the start of the section. But we have some respect for the complexity of the processes required.

It is tempting to say that the probability of life forming spontaneously is almost zero. When we think of the formation of DNA, the different types of RNA, the ribosomes, the proteins, and the complex structures of each, it is reasonable to doubt that this could ever happen more than once. This is in fact the view of many biologists, who say that, if life on Earth was not transported here (i.e., by the panspermia theory), then its occurrence is such unbelievable luck that it could never happen again.

But there is a counterargument. Life on Earth formed within one billion years after Earth was created. On the cosmic scale, this is not a lot of time. Laboratory experiments have shown that amino acids can be created from the known components of the primeval Earth's atmosphere, and from energy supplied by lightning or volcanoes. And we have found amino acids on comets and asteroids. The construction of life's building blocks seems easy and reproducible on other planetary bodies.

The fact that amino acids seem to be easy to produce, coupled with the short amount of time needed for life to develop on Earth, leads many to believe that the spontaneous generation of life is not such a long shot. Thus, these scientists estimate f_l to be 0.5 or greater. Some believe the emergence of life is inevitable on habitable worlds and put the likelihood close to 1.

I think an estimate of 0.5 is extremely optimistic. In our solar system, if we think of Venus, Earth, and Mars as potentially habitable, life only developed on one of the three. This gives f_l for our solar system as only 0.33, and this is surely a biased statistic, since it includes our Earth. I will say that the probability of life developing spontaneously on a habitable planet is about 1 in 20, or 0.05.

THE CHANCE OF INTELLIGENT LIFE DEVELOPING

Given that life exists on a planet, what are the chances that the life can eventually evolve sufficiently to produce intelligent beings? This is the quantity f_i in the Drake equation. Recall from our definition of intelligence that intelligent beings must be capable of communicating through space.

The development of intelligence is a tall order. All that we require to start is some primitive bacteria or a rough equivalent, and we are asking for the eventual existence of beings who can travel or send signals to the stars. It has happened here on Earth. How lucky were we?

The Requirements for Intelligence

Beings must have certain attributes to become intelligent. We can try to make a list of such attributes.

The most obvious is brain size. Organisms that are going to forge an advanced technology must have large brains. Technological advances require creative and imaginative thought processes. It is helpful to reflect on this for a second. Many devices we take for granted—for example, the wheel or the internal combustion engine —may appear simple to us once we understand them. But it is a far different thing to understand someone else's creativity after the fact than it is to invent a device. These inventions required an enormous amount of creativity to develop. If you consider all the inventions that have taken place in human history in order for us to attain our current technology, you begin to appreciate the importance of brain size.

But brain size clearly is not enough. Dolphins, elephants, whales, and polar bears all have large brains. But none of them will ever be able to send signals to the stars. Why?

Tool-making ability is probably the next most important requirement. Tools and machinery are necessary for beings to develop an advanced technology.

Of course we can imagine advanced beings, similar in nature to our electric eel, that can generate electric fields. Perhaps such beings exist in the galaxy; beings that can communicate through space by generating alternating electro-magnetic fields, thus sending signals through space. Such beings are certainly possible in theory, but questions must be asked. How would such beings learn about the cosmos without tools with which to look at the heavens? How would they figure out that theirs is just one tiny world orbiting an insignificant sun? Without this knowledge, why would they be interested in sending signals into space?

It was only by our use of tools that we were able to determine the nature of the stars. Before the invention of tools, humans had no idea what the stars were, or even the moon. Tool-making ability is a necessary skill toward the development of intelligence.

Tool-making ability implies other requirements. An organism must have free limbs that it can use to create the tools. A polar bear has no free limbs, because it needs all four limbs for transportation. Therefore, a polar bear would probably never be able to build tools of any sophistication.

An organism needs to be land-based, because only on the land are sufficient metals available for building sophisticated tools. Metals such as iron, tin, and nickel are just not found floating through liquids. Human beings would certainly have no technology without metals available for tool building. Dolphins and whales probably have sufficient intelligence to build tools, but, unfortunately, they are water-based (not to mention that they have no free limbs).

Finally, an organism must feel a pressing need to build tools. This is important. If an organism can go about its business of survival quite happily, then why should it bother to build tools? Why should it expend its time and energies on such a vague pursuit?

Human beings had a pressing need to build tools. Tool-building ability gave man the advantage over stronger and swifter wild beasts. If Earth had not been full of wild and dangerous beasts, perhaps the early humans would never have built tools. Mankind might have been content to spend its days in relative calm, gathering berries and drinking water from streams.

The pressing need to build tools must come from the planet's ecology. A rich, diverse set of life forms seems to be an essential requirement. Of course, we can probably imagine other scenarios that would provide creatures with a pressing need to build tools. But it is unlikely that a burning desire for knowledge of the universe would provide that pressing need, at least at the start.

Communication at an advanced level is also critically important to intelligence. Intelligent beings must possess an advanced language in order to communicate ideas to each other, and to pass ideas down to later generations. Humans possess the ability to create a wide variety of sounds, thanks to vocal cords, teeth, and a flexible, easily controlled tongue.

Other animals are not so fortunate. While apes could probably develop a complex language based on hand signals, it would be an enormous creative leap for an ape or a group of apes to get this idea on their own. The first instinct of creatures (at least on Earth) is to communicate through sounds. If one creature wants to communicate to another the idea, "I am mad at you," the right set of sounds communicates this quite effectively. Hand signals in themselves have no accompanying emotional weight. Creatures begin communicating by conveying emotions. This is why, at least on Earth, almost all communcication is by sound.

Of course, alien beings could develop ways to communicate with each other that are undreamt of by us. But a necessary requirement is that the communication language be very sophisticated. Such a language must be intricate enough to express such ideas as "In a right triangle, the square of the hypotenuse equals the sum of the squares of the other two sides," or "Atoms heavier than iron won't create energy by fusing together."

Still another requirement is life span. Intelligent beings that develop a technology must hand down their knowledge from generation to generation. Newly born members of the species must spend a good portion of their lives learning the existing technology developed by their forefathers. Having accomplished that, they must then spend a portion of their lives advancing that technology.

Training takes time, and can only be done if beings have a sufficiently long life span. Again, humans are fortunate in this regard. Their lives are long enough to allow for the growth and advancement of technological ideas through generations.

Evolution—A View from the Top Down

The concept of evolution, once considered heresy, is now a well-respected and generally accepted theory among scientists. In a broad sense, the theory says that simple life evolves over time into more complicated life by a process known as *natural selection.*

Another way to express this idea is through the often used phrase "survival of the fittest." The idea goes like this. Not all beings are created equal. All possess subtle nuances that set them apart from others. As hard as we look, we will never find any two creatures exactly alike.

Some creatures have slight differences that enable them to survive easier and to reproduce more offspring. Thus, over time, more offspring are produced by the creatures that are the "fittest," or that, because of subtle unique qualities, are favored over others.

Taken over the course of several generations, creatures that are not well adapted to survival become extinct. Others prosper. The members of a single species that are better adapted than others in that species are healthier and produce more offspring, and the species advances over time.

Over the course of billions of years, life on Earth evolved from bacteria and one-celled primitive organisms to its current state. It accomplished this through the long process of evolution.

Slowly, subtle changes in one-celled animals occur, thus producing multi-celled animals, and so on. Many times, these subtle changes, called *mutations*, work to an organism's disadvantage. But, rarely, a chance mutation leads to an organism that is better adapted to survival than its counterparts. The mutant produces offspring, which also prosper, sometimes at the expense of the more primitive creatures. The very slow process, driven by chance mutations, leads to higher and higher forms of life. Human beings, with their complex brains, upright posture, free limbs, fingers, vocal cords, flexible tongues, and all else, developed from such a series of chance mutations over the course of billions of years.

Evolution—A View from the Bottom Up

Mutations occur because of errors in the copying process as DNA replicates itself. This insight came from discoveries that certain genetic diseases, such as sickle-cell anemia and Huntington's chorea, occur because of a single mistake in one codon of one gene. Because of this mistake, an incorrect protein is built, based on a single incorrect amino acid in the chain.

Sickle-cell anemia and Huntington's chorea are types of mutations. They are random changes in the genetic code, which, like most such random changes, are not for the better.

A mutation that works for the benefit of an organism is an extremely rare event when viewed from the context of the molecular level. Such a mutation would have to change randomly an amino acid of a protein chain, creating a new protein as a result, and the new protein would have to be better than the old protein.

The unlikelihood of such a favorable alteration can be appreciated by considering the comparison between proteins and English sentences. If an amino acid (of which twenty are used) is represented by a letter of the alphabet, then a peptide chain can be represented by a word, and a protein by a sentence. Consider a complicated sentence, appropriately representative of a complex protein in a human being, and consider the chances of altering one (or maybe two at most) letters in the sentence and actually improving the sentence by doing so. It is very unlikely.

Why mutation-causing DNA copying errors occur is not completely understood. One possibility is that DNA can be damaged by radiation, such as X-rays, cosmic rays, or ultraviolet rays. Another is that an inadequate diet may be a culprit, as adequate nutrition is necessary to maintain a healthy cell chemistry and to combat the invasion or growth of potential DNA-damaging toxins.

Summary

Opinion is divided on the likelihood of intelligent life developing on a planet where primitive life already exists. Astronomers and physicists believe that the development of intelligent life is fairly inevitable over time.

Molecular biologists disagree. The different paths that can be taken by subsequent mutations are so many, with only an insignificant handful leading to intelligent life (again, meaning that it is able to communicate across interstellar space). On Earth, we have had this experience, where only one species out of billions can be considered intelligent.

There is perhaps even more luck involved than the occurrence of the correct chancy sequence of genetic mutations. Mankind's pressing need to develop tools only was realized because he was threatened by beasts. If mankind had also been stronger and swifter than the other creatures in his ecology, he might never have felt a need to develop tools.

Man's brain developed, to a large extent, because of stress induced on him by a difficult environment. Not only the presence of dangerous animals, but also climatic changes brought on by ice ages or volcanic eruptions, put stress on mankind, forcing him to think. Chance mutations that led to increased brain size became favorable. Man's timing was also somewhat important. He did not arrive on Earth at a time when fearsome dinosaurs roamed. If he had, he might have become extinct very quickly. Man's development did not encounter cataclysm, such as impacts of giant asteroids or supermassive volcanic eruptions. Man encountered just enough stress to accelerate his evolution, but not enough to destroy him. It was a fortunate scenario indeed.

Many molecular biologists believe that intelligence exists nowhere else in the galaxy. Optimists believe it is common and a fairly inevitable development. These optimists set the value of f_i to about 1 in 100. For reasons cited throughout this section, I will use a less optimistic number. I will say it is 1 in 10,000, or 0.0001.

THE LIFETIME OF AN INTELLIGENT SPECIES

Mankind began sending radio waves into space roughly 100 years ago. Thus, by our definition, man has been intelligent for about 100 years. How long will he remain intelligent before he either becomes extinct, loses his intelligence, or destroys himself? This number, in years, is the quantity L in the Drake equation.

This is important, because our chances of communicating with another intelligence depend on our timing. An alien intelligence trying to communicate with us would, so far, only have had 100 years out of 4.5 billion (the age of Earth) during which we could communicate.

There are many ways that man could become extinct. He could run out of natural resources that supply him with energy. He could die because of excessive pollution or self-induced climatic change. He could create any number of ecological disasters, destroying green plants, for example. He could die from natural cataclysms such as ice ages, massive volcanic eruptions, or asteroid impacts. Or he could kill himself off in a global war.

It is natural that man's ability to destroy himself came shortly after his arrival at "intelligence." Thus, shortly after man invented radio waves, he invented atomic energy. Each creature that arrives at intelligence probably faces the same dilemma. They all then must constantly dance to the tune of avoiding self-destruction.

Some mathematically oriented sociologists have attempted to estimate the expected remaining lifetime of humanity. They have assumed that the biggest danger confronting man is global war, and have constructed mathematical models based on the history and frequency of wars.

In these models, wars are assumed to be historically random events. In other words, wars are assumed to occur at random points in time, and the occurrence of a war does not depend on the history of past wars or on the time interval since the most recent war. According to these models, the history of warfare is only important in estimating the average number of wars over time.

An analogy might be watching cars on a lonely country road. You might estimate that twenty cars per day pass down the road. If you go to the side of the road and sit, how long are you likely to have to wait until you see one? Such a model makes use of the exponential probability distribution. (For an explanation of the exponential probability distribution, refer to *Becoming a Mental Math Wizard* by Jerry Lucas, also published by Shoe Tree Press.)

Wars present a more complicated situation than cars, since they come in vastly different sizes. But based on past human history, different mathematical estimates for the remaining lifetime of man have ranged from 200 to a few thousand years.

The models have not received much publicity, and this is understandable. It is not at all clear that the future history of mankind can be predicted by mathematics. Over the centuries, human beings have proved to be remarkably resilient and tenacious in finding ways to survive as a species. Mathematical models do not capture these aspects of the human spirit.

I am more optimistic about mankind's ability to survive as a species, and thus for other similar species to survive, once they acquire an advanced technology. I will use a value of L of 100,000 years, the average lifetime of a species, once it has become intelligent.

WAIT A MINUTE! WHAT ABOUT UFO'S?

There are those who would disagree with this entire chapter. They would claim that there is alien intelligent life, and we know about it. The proof is in the unexplained UFO sightings. We don't need to use all this analysis and Drake's equation and everything else. We already know the answer. Intelligent extraterrestrial beings exist.

The scientific community does not accept UFO's as proof of intelligent life on other worlds. Many of the UFO sightings can be exposed as honest mistakes or hoaxes. But even the ones that cannot be explained do not prove the existence of extraterrestrials.

The key idea here is the scientific principle called Occam's Razor. According to Occam's Razor, when an observation can be explained by two or more competing theories, science requires us to accept the simplest theory, or, in other words, the one most likely to be true.

Here is how Occam's Razor works. Let's suppose I am somebody you know. I have just returned from a vacation. I tell you the following story.

"I went to Florida on my vacation. I was swimming in a river and was attacked by an alligator. The alligator tried to drown me by holding me underwater. I pretended to be dead. The alligator thought I was dead, so he dragged me along through the water to his home along the edge of the river. He dropped me on the ground, then walked a few feet away, lay down, and fell asleep. I got up and tried to sneak away, but I woke up the alligator. He came at me. I picked up a rock and stabbed him in the throat, killing him. Then I got away."

Do you believe this story or don't you? Perhaps I am your friend, and you don't want to think I would lie to you. But if you are a scientist, in pursuit of the truth, you must maintain complete objectivity. How do you decide what to believe?

Your observation is that I have told you a story. You can now form two competing theories to explain this observation. The first theory is that I am lying. The second theory is that I am telling the truth.

What is the likelihood that I am lying? You may know me as a trustworthy friend, not likely to tell a lie. But it is possible.

What is the likelihood I am telling the truth? There are five billion people in the world, and very few of them, during the course of an entire lifetime, have an experience such as this. I have experienced it just last week. The likelihood of this is probably something like one in a million.

Based on the principle of Occam's Razor, you as an objective scientist should conclude that I am lying. It is the most likely theory that explains what you have observed. Perhaps I don't tell many lies, and any one thing that I say is probably not a lie. So it is unlikely I would tell a lie. But the alternate theory, that I am telling the truth, is even more unlikely.

Now suppose I give you some evidence. I show you my plane tickets to Florida. I show you a large wound on my shoulder. You study the wound, and conclude that it is possible it could have been caused by an alligator bite. Now do you believe me?

Probably not. The likelihood that I am lying is changed by the evidence. But it is still much more likely than the alternative. You don't know how I got that wound on my shoulder. You cannot offer any explanation for it. But that doesn't prove I am telling the truth. It just means that you don't have a way to explain what I am showing you. You are not admitting that other explanations are impossible, you just don't have one to offer.

Now, if we both went to Florida, found the spot on the river where I was attacked, located the dead alligator where I said it would be, performed an autopsy on the alligator, verifying how and when it died, and everything was consistent with my story, then you would probably believe me.

The point is this: fantastic claims must be backed up by fantastic evidence. Otherwise, we are compelled by Occam's Razor to disbelieve the claim. UFO sightings represent fantastic claims that are not backed up by fantastic evidence. And just because some have not been explained does not compel us to believe them.

PUTTING IT ALL TOGETHER

We are ready to combine all our numbers from the Drake equation and come up with our estimate for the number of planets in the Milky Way Galaxy that have intelligent life.

We can group together all the parameters that we have estimated so far:

$R_* = 0.2$; $f_p = 0.5$; $n_e = 3$; $f_l = 0.05$; $f_i = 0.0001$; $L = 100,000$

We multiply all these quantities together to get the expected number of planets in the Milky Way Galaxy that are the home to intelligent life. When we do this, we come up with a result of 0.15. The result is extremely significant, since it is considerably less than 1.

There are many people more learned than myself who would disagree with me here. In fact, most of the numerical estimates of the Drake equation quantities given in this chapter, due to scanty information, are based on guesswork.

Nonetheless, I have laid out as objectively as possible the problem as I see it. Based on the best information we currently have available, for what it is worth, I will tell you what I think. I think that there is an excellent chance we are the most advanced form of life in the Milky Way Galaxy right now.

BIBLIOGRAPHY

Books

Bova, B., and Preiss, B., editors. *First Contact: The Search for Extraterrestrial Intelligence*. New York: Penguin Books, 1991.

Cairns-Smith. *Seven Clues to the Origin of Life*. Cambridge: Cambridge University Press, 1985.

Calvin, W. *The Ascent of Mind*. New York: Bantam Books, 1991.

Davoust, E. *The Cosmic Water Hole*. Cambridge: MIT Press, 1991.

Dobzhansky, T. *Genetic Diversity and Human Equality*. New York: Basic Books, 1973.

Drake, F. *Intelligent Life in Space*. New York: Macmillan, 1960.

Edey, M., and Johanson, D. *Blueprints: Solving the Mystery of Evolution*. New York: Penguin Books, 1990.

Editors of Time-Life Books. *Life Search*. Alexandria, VA: Time-Life Books, 1988.

——. *Stars*. Alexandria, VA: Time-Life Books, 1988.

Gardner, M. *The New Ambidextrous Universe*. New York: W.H. Freeman and Company, 1990.

Goldsmith, D., and Cohen, N. *Mysteries of the Milky Way*. Chicago: Contemporary Books, 1991.

Gribbin, J. *In Search of the Double Helix*. New York: Bantam Books, 1987.

Lucas, J. *Becoming a Mental Math Wizard*. White Hall, VA: Shoe Tree Press, 1991.

Regis, E., editor. *Extraterrestrials: Science and Alien Intelligence*.

Cambridge: Cambridge University Press, 1985.

Schroedinger, E. *What is Life?* Cambridge: Cambridge University Press, 1944.

Seielstad, G. *At the Heart of the Web: The Inevitable Genesis of Intelligent Life*. Orlando, FL: Harcourt Brace Jovanovich Inc., 1989.

Periodicals

Black, D. "Worlds Around Other Stars." *Scientific American,* January 1991.

Campbell, B. "Planets Around Other Stars." *The Planetary Report*, May-June 1988.

Crick, F. "The Seeds of Life." *Discover*, October 1981.

Croswell, K. "Does Alpha Centauri Have Intelligent Life?" *Astronomy*, April 1991.

Dickerson, R. "Chemical Evolution and the Origin of Life." *Scientific American*, September 1978.

Hellerstein, D. "Plotting a Theory of the Brain." *New York Times Magazine*, May 22, 1988.

McKean, K. "Life on a Young Planet." *Discover*, March 1983.

Olson, E. "Intelligent Life in Space." *Astronomy*, July 1985.

Papagiannis, M. "Bioastronomy: The Search for Extraterrestrial Life." *Sky and Telescope*, June 1984.

Woese, C. "Archaebacteria." *Scientific American*, June 1981.

Epilogue:
What Moves Science Forward?

Three hundred and some years ago, there lived a shy, awkward, clumsy man named Isaac Newton. He seemed to have a bizarre fascination with understanding what caused objects to fall to Earth.

In the process of persistently pursuing this mystery, Isaac discovered the universal law of gravitation, invented the branch of mathematics called calculus, and explained the physical world with a brilliance of clarity and vision.

When human beings landed on the moon centuries later, the achievement was described as a team effort. In a real way, Sir Isaac Newton was the star player of that team, for it was his ideas and his equations that were programmed into NASA's computers.

Yet Newton spent his entire professional life shrouded in controversy. His ideas so challenged the conventional ways of thinking in his day that they were not easily accepted.

Alfred Wegener, who put forth the theory of plate tectonics, was ridiculed severely for his bold challenge of conventional wisdom. He died in 1930, on an expedition to Greenland, without the slightest hint that his ideas would one day form the basis of modern geology.

And so it was with Galileo, Copernicus, Darwin, and even Einstein. All were individuals who put forth bold ideas that challenged the establishment. They all died not realizing the mighty effect that their ideas would one day have on the scientific community.

If they had one thing in common, it was their relentless and tireless persistence in finding the answer to a mystery. They were not afraid of ridicule. Their only goal was to understand the truth. And to these individuals, and to the countless others who have advanced scientific knowledge, the world owes an enormous debt of gratitude.

The force that moves science forward does not come from corporations with large research budgets or from federal bureaucracies. It comes from deep within the individual. It comes from relentless persistence in solving troubling mysteries, and from the courage in challenging established thinking. The force that moves science forward is the strength of the human spirit.

Index